Journey Toward Energy Independence

Harry Johnson

Harry R. Johnson (email: harryjohnson35@hotmail.com)

"Journey Toward Energy Independence," by Harry Johnson. ISBN 978-1-62137-987-4 (softcover); 978-1-62137-988-1 (hardcover).

Published 2017 by Virtualbookworm.com Publishing Inc., P.O. Box 9949, College Station, TX , 77842, US.

©2017 Harry Johnson. All rights reserved. No part of this publication may be reproduced, stored in a retrieval system, or transmitted in any form or by any means, electronic, mechanical, recording or otherwise, without the prior written permission of Harry Johnson.

Contents

Introduction 1

Chapter 1 3

  Waterflooding Applachian Oil Fields 3

    Predictive Models 5

    Statistics And Regression Models 8

Chapter 2 11

  Navy Dumps Coal And Switches To Oil 11

    Oil Shale Resources 12

    Oil Shale Retorting Technology 13

Chapter 3 17

  The Petroleum Industry Of The 1960s 17

    Automobiles And Air Pollution 18

    Higher Education 19

Chapter 4 23

  Stimulating Oil Shale Development By Leasing 23

    Prototype Leasing Program Gets Organized 25

    Environmental Impacts Of Oil Shale Development 26

    Environmental Impact Statement Gets Reviewed 31

    Prototype Lease Sales Win Large Bonus Payments 32

Chapter 5 35

  United States Responds To The Saudi Arabia Oil Embargo 35

    Project Independence 36

    Energy Research And Development Administration 38

    Synfuels Fuels Commercialization Program 39

Chapter 6 43
Clash Of Government Energy Plans 43
Energy Research And Development Administration Energy Plans 44
Federal Energy Agency Plans 46
Natural Gas Politics In The 1970s 48
Senate Hearings 49
Briefing President Jimmy Carter 50
House Of Representative Hearings 51

Chapter 7 59
Department Of Energy Created 59
Strategy For Energy Technologies 59
Doe Created, But What About George? 62
Project Implementation Moves To Field Locations 62

Chapter 8 65
United States Responds To The Iran Civil War 65
Massive Synfuels Program 67
Enhanced Oil Recovery For Conventional Oil 69
Oil Recovery Engineering 70
Arab Response To The U.S. Actions 74
Lasting Memories Of This Volatile Period 80

Chapter 9 83
Unconventional Natural Gas Production 83
Deposited, Mashed, Heated, And Fractured 83
Migrated, Trapped, And Surface Seeps 85
Fracture Science 89
From Fracture Science To Fracture Engineering 91
From Fracture Engineering To Research Horizontal Wells 94
From Research Horizontal Wells To Commercial Horizontal Wells 95
Water Management In An Environmental Age 99

Chapter 10 103
　Unconventional Oil Production 103
　　Initial Focus On Oil Shale 104
　　New Massive Synfuels Program Being Developed 106
　　Horizontal Wells To The Rescue... 106
Chapter 11 111
　United States Responds To The Libyan Civil War 111
　　Strategic Petroleum Reserve Crude Compatibility Evaluated 113
　　Strategic Petroleum Reserve 2011 Libyan Drawdown Effective 114
Chapter 12 117
　Completing The Journey Toward Energy Independence 117
　　Flow Of Oil Imports Since 1960 117
　　Effective Response To Oil Import Disruptions Developed 119
　　Massive Unconventional Shale Deposits... 121
　　More Oil More Water 124
Acknowledgements 127
References 131

Figures

Fig 1. Government Incentives Increase U.S. Oil Production 76

Fig. 2. Cheap Foreign Oil Forces U.S. Oil Production Lower 79

Fig. 3 Vertical Fractures in Marcellus Shale Core 85

Fig. 4 Twenty-one Oil and Gas Traps 87

Fig. 5 Drill Horizontal Wells to Intersect Natural Fractures 91

Fig. 6 Active DOE Eastern Gas Shales Projects in 1978 93

Fig. 7 United States Gas Production Reaches All-Time High 98

Fig. 8 United States Oil Imports Reach Record High in 2004 103

Fig. 9. Unconventional Oil Production Revives U.S. Petroleum Industry 107

Fig 10. Massive Unconventional Shale Deposits Widely Dispersed Across U.S. 121

Introduction

LITTLE DID I KNOW in the beginning that I was about to start a journey that would involve an oil embargo, two civil wars, oil supply gluts, friendly and not so friendly encounters with the U.S. Congress, and an energy briefing of the President in his Cabinet Room.

Long lines at the gas pumps caused by the Arab oil embargo of 1973 served notice that the United States was no longer in charge of its energy future. The nation was still not prepared in 1978 when oil imports were interrupted by a civil war in Iran. This time, long gasoline lines were accompanied by domestic riots and panic at the pumps. The government had to act, and it did with regulations, tax incentives, executive orders, and accelerated research funding that significantly increased domestic oil production. Oil exporting nations saw increased oil production as a threat to their market share and flooded the world market with cheap oil. Many U.S. producers went bankrupt, banks failed, and domestic oil production fell to a 60-year low.

Chapters four through eight chronicle the volatile 1970 period, the actions of oil exporters, and the reaction of the U.S. government to these actions. The opening chapters trace my early career that helped prepare me to support government programs formed during this period.

Two events since the 1970s have made the nation more energy secure and on the road to energy independence.

First, the petroleum industry developed horizontal well drilling and massive hydraulic fracturing to recover oil and gas from dense unconventional resources. Sustained government research described in chapters nine and ten provided the foundation for industry interest and development of the horizontal well drilling technology.

Second, the Strategic Petroleum Reserve was full and ready when the 2011 civil war in Libya caused a worldwide oil shortfall, described in chapter eleven. The U.S. delivered 30 million barrels of oil as part of the international effort to mitigate the shortfall impact. The drawdown worked as planned without incident. No long lines or panic at the pumps this time. The U.S. and the world had developed an effective shield against supply interruptions.

The United States has already achieved a high level of energy security that was not available in the 1970s. And, as described in chapter twelve, shale deposits are widely distributed within the United States and contain decades of oil and gas supply. These vast shale resources coupled with horizontal technology will likely make the U.S. energy independent within the next decade.

Chapter 1

WATERFLOODING APPLACHIAN OIL FIELDS

IT WAS AFTER MIDNIGHT and I was stumbling through the cavernous, darkly-lighted, computer center at West Virginia University, looking for paper. Not just any paper, but the cardboard paper needed to feed a waiting monster known as the IBM 1620 computer. Dean Boley and I were debugging a large mathematical model we had developed, and the University allowed us after-hours access to the largest computer then available.

I found a box of cardboard paper and returned in the dim light to the IBM machine. As we loaded the paper into the machine, Dean noted unusual notches in the cards being spit out of the computer. We found that we were keypunching cards that had already been punched for another project.

This was the start of a number of apologies to the University for trying to screw up a PhD research project. But it was also an important milestone of our Bureau of Mines project to help local producers increase oil production in the Appalachian regions of Pennsylvania, Ohio, and West Virginia.

The Drake well, drilled in northern Pennsylvania in 1859, sparked the first oil boom in the United States. Drilling spread rapidly in the Appalachian region as reservoirs were quickly discovered and developed. Oil and gas are initially trapped in the reservoir rock. When the trap is penetrated by drilling a vertical well, the trapped natural gas expands and drives oil to the well bore for production. This solution-gas drive mechanism can rarely produce more than 25 percent of the original oil in place, leaving behind 75 percent of the oil that was found. Additional energy was needed to improve recovery of the remaining high-quality crude.

In 1959, one hundred years after the Drake well, Appalachian oil producers contacted the United States Bureau of Mines to determine if the Bureau would consider a project to improve oil production from pressure-depleted reservoirs. In addition to active industry participation, the decision to proceed required the collection of an enormous amount of field data that was not then available. Once collected, the field data needed to be incorporated into a mathematical model or models to predict oil recovery over time due to the injection of gas or water into the depleted reservoir. Of course, mathematical models for Appalachian oil reservoirs did not then exist.

The Bureau initiated this project in 1960. Congress allocated funds and construction of an oil laboratory started at the Bureau's Coal Research Center located in Morgantown, West Virginia. Cooperative agreements with industry partners were negotiated and signed. Hiring of staff to carry out the project was initiated. I had just graduated as a petroleum and natural gas engineer from the University of Pittsburgh and was selected to support the laboratory development and use of the laboratory results to model reservoir performance.

Predictive Models

Universities have laboratories designed to teach petroleum basics. University laboratories are not designed to measure technical data in the detail needed to feed a sophisticated model used to predict waterflood performance. At the Bureau, equipment was purchased, or constructed, and installed to measure the physical properties of the reservoir, its fluid properties (oil, water, and gas), and how the oil, gas, and water move in the reservoir. All of the field information was received at the Bureau laboratory for analysis, and the data used in our predictive models.

Back in the dark ages of the early '60s, the Microsoft suite of products to manage large amounts of numerical data had not yet been invented. Scientific programming used FORTRAN, which was developed by IBM technicians in the 1950s. By 1960, versions of the new programming tools were available for the IBM 1620 computer.

Modeling begins by constructing a very detailed flow diagram to represent the flow of data in the predictive model. Once the model flow was documented, it was programmed using primitive codes which were then being developed by IBM. Data flow in the math model could easily be captured and the data circulated until the problem was found and corrected. In short, it took months for us to program and debug the first predictive model for use in this project.

Waterflood Possibilities of the Clinton Sand, Logan Oilfield, Hocking County, Ohio by Harry R. Johnson and Dean W. Boley was the first technical report prepared by Bureau authors, and it appeared in Producers Monthly magazine, December 1963. The article summarized the technical data collected for the project.

Two different calculation methods were used to predict waterflood performance "...and were programmed by the authors for computer (IBM 1620) solution." We found that the Clinton sand at this location contained a sufficient quantity of mobile oil and enough water injection capacity for an economic waterflood.

Waterflooding requires a minimum of five wells. The production well is surrounded by four wells drilled on a pattern. Water is injected into the four wells, and the injected water forces mobile oil to the center producer. The initial pattern is replicated until the field is fully developed for waterflood operations.

Bureau engineers analyzed other important oilfields located throughout the Appalachian area:

- Kane Oilfield, Elk County, PA in 1964;
- Burning Springs Pool, Wirt County, WV in 1964;
- Logan Oilfield, Hocking County, Ohio in 1965;
- Bonds Creek Oilfield, Pleasants County, WV in 1966;
- Sartwell Oilfield, McKean County, PA in 1967; and
- Kane Oilfield, Elk County, PA in 1967.

Theoretical and Field Waterflood Performance, Kane Sand, Kane Oilfield, Elk County, PA by Leo A. Schrider, John R. Duda, and Harry R. Johnson, Bureau of Mines Report of Investigations 6917, published in 1967, showed that oil production from a 36-acre pilot waterflood was greater than originally forecast. Assumptions used in the predictive model were updated to better match model performance with actual results. More importantly, the favorable field response prompted the operator to expand his pilot project from 36 acres to 180 acres. Expected oil production from the 216-acre project totaled over 300,000 barrels, and this oil would have been left behind if a waterflood had not been used.

Bureau analyses and favorable field performance demonstrated that the pressure-depleted fields of the Appalachian area would respond to water injection to recover oil. Moreover, the response was predictable, and this permitted a reasonable estimate of project economics. Over time, Appalachian oil producers adopted waterflooding as a standard approach to increase oil recovery from depleted reservoirs, probably by millions of barrels.

The project objective to improve oil production from Appalachian pressure-depleted fields had been achieved. Bureau researchers then turned to more advanced efforts to better characterize for development the Devonian shales that underlay most of Pennsylvania, West Virginia, Ohio, and New York. The shales were known to contain oil and gas, but they did not have the capacity to flow liquids from the formation to the vertical wellbore. The shales were not of commercial interest for development using vertical well technology.

A seminal paper published in 1968 by my friend and geologist Bill Overbey showed how to predict the orientation of the natural fractures in the Appalachian area. Given that knowledge, another friend and fellow petroleum engineer, Joe Pasini, suggested that we drill horizontal wells perpendicular to the fractures to intersect as many of the fractures as possible, to increase oil and gas production. In 1976, Pasini and Overbey were awarded the first horizontal well patent based on this work.

This early government research provided the technical foundation for industry interest and the development of horizontal well drilling and hydraulic fracturing. Application of the new technology by industry began in 1985 in Texas, and spread rapidly to the Marcellus shale in the east, the Bakken shale in the west, and to other shale locations. The nation's oil and gas production increased rapidly, reversing a 40-year decline in petroleum production from conventional vertical wells and leading the nation toward energy independence.

The formation of the shales some 400 million ago, the creation of natural fractures in the shale, and the ability to predict and use the fractures to produce oil and gas is detailed in chapter 9 in this narrative. Readers who jump to this chapter will miss 40 years of our growing dependence on oil imports, supply interruptions by embargo and civil wars, supply surges that nearly killed the domestic oil industry, and the nation's response to these events.

For now, I am returning to the chronological description of the journey that put me near the center of the nation's efforts to manage supply shortfalls caused by oil imports.

Statistics and Regression Models

Leo Schrider was a year behind me in Petroleum Engineering at the University of Pittsburgh. Following graduation, he went to work for Shell in New Orleans. However, Leo found that his dad was terminally ill and wanted to return to the Pittsburgh area to support his family. I hired Leo for our waterflood work and, over time, he began to take over my project management responsibilities. He and his wife Jane also become reliable camping buddies with our family.

My focus then shifted to in-situ oil recovery, a very advanced oil recovery technology that had been tested in the Appalachian area without economic success. The issue was how to get enough carbon deposited near the well bore to start and sustain combustion in order to create the energy needed to force oil to a well for production.

I needed a variety of technical capabilities to address in-situ combustion. Oil chemistry was provided by Ed Burwell, an experienced Bureau chemist. Statistical guidance was provided by John Holden, a chemical engineer in the Bureau's coal research center. Together, we designed and conducted laboratory experiments that showed enough carbon could be deposited under the right time and temperature conditions.

Carbon Deposition for Thermal Recovery of Petroleum; A Statistical Approach to Research by Harry R. Johnson and Edward L. Burwell was prepared as a Bureau report and reprinted by Producers Monthly, July 1966. The work led to requests for me to teach the statistical approaches used in our research. It also led to a deep understanding of how to prepare and use a mathematical model to represent a physical system, a skill that would prove to be very useful in my next assignment.

The Bureau maintained a very high standard before a technical report was approved for publication. Experts reviewed each draft publication and technical issues were resolved. Washington Bureau staff reviewed the final draft and any remaining issues were resolved. By the mid-60s, I had prepared and published seven highly technical publications. Results were presented at public meetings. The Washington staff must have liked what they saw, and I was offered a promotion to D.C. in 1964.

The decision to uproot our comfortable life in Morgantown and move to the hectic D.C. area was made with my wonderful wife Louise, now a partner of over five decades and counting. We accepted the promotion, packed the house and our two hillbilly kids—Leslie, age three, and Ray, age three months—and found our first home in suburban Maryland. No room to pack our friends and co-workers, but as it turned out, I was able to recruit many of them to help share adventures that followed.

Chapter 2

NAVY DUMPS COAL AND SWITCHES TO OIL

IT WAS ABOUT 1870, and a Navy fighting ship pulled into port on the west coast of the U.S. and docked. Sailors then began to load the ship with coal from land storage. Once full, the ship steamed out to sea, burning coal to raise steam for propulsion. Imagine the coal being fed to combustion boilers and the ash removed for sea burial. Burning coal was a messy, labor-intensive operation.

Drake's 1869 discovery of oil and the rush for development changed Navy operations. Oil only required a hose coupled to the ship to quickly transfer the liquid fuel from land to sea storage. The oil also burned far cleaner than coal. The Navy began to convert all of its fighting ships from coal to oil. By 1910, the increasing demand for oil led the Navy to question whether the supply was adequate in the event of war or a national emergency.

Congress responded to these concerns and authorized the President to withdraw large areas of potential oil-bearing land from commercial development in California, Wyoming, Colorado, and Alaska. The Department of Interior assumed responsibility for land management, and its Bureau of Mines assumed responsibility for energy technology development.

The Bureau established an oil and gas technology center in Bartlesville, Oklahoma, and an oil shale technology center in Laramie, Wyoming. By 1920, research on how to liberate oil from oil shale had begun at Laramie and at the Naval Oil Shale Reserves at Anvil Points, Colorado.

This field research was still underway when I arrived in 1964 for my new assignment at the Bureau's Washington headquarters.

One of the first questions I was asked: Can you model the oil shale process and help optimize oil recovery? Where is the oil shale, I asked? Can we waterflood it, I wondered?

And that's how I found myself on a narrow road up a mountain in Colorado to view the oil shale mine the Bureau had been developing for nearly five decades.

Oil Shale Resources

As it turned out, oil shale is an organic-rich deposit where the organic matter had not been converted to liquid by heat and pressure. In the vicinity of the Bureau's mine, it is a semi-solid that crops out several thousand feet above the valley floor at Anvil Points, Colorado. From the outcrop, it dips to the north and increases in thickness. At its deepest point, the oil shale becomes nearly 2,000 feet thick. The oil in the Colorado deposits exceeds one trillion barrels, making it the largest organic deposit in the world. No wonder the Navy wanted oil shale as a supply backup.

The Bureau's oil shale mine was impressive: a ceiling over 70 feet high, with rooms 60 feet by 60 feet. The ceiling was supported by large pillars left behind in the mining process. Raw mined oil shale was loaded on large trucks and carried down the winding, narrow mountain road to the valley floor. The trucks always had the lane closer to the mountain. We met one coming up the mountain when we were going down in the car. I got a good look from the passenger seat of the car at the magnificent valley. Then I looked straight down and I became terrified—there was no guard rail, only a direct drop to the valley! This was my first and last visit to the mine.

Once we returned to the safety of the valley, I was introduced to the machine used to process the mined oil shale, converting it to oil.

Oil Shale Retorting Technology

Commercial oil shale technology had its start in Scotland about 1860. The Scots developed a vertical kiln technology that was adopted and improved by Bureau of Mines engineers at the Anvil Points facility. The kiln can be described as a large diameter pipe erected vertically. Gates were welded on the top of kiln to control the introduction of shale, and gates at the bottom controlled its exit speed. Welded internal workings included a combustion chamber and a way to collect the oil liberated from the shale.

The process is relatively straightforward:

- Crush the shale to a proper size;
- introduce the sized shale into the top of the vertical kiln;
- flow the shale by gravity from the top to the bottom of the kiln;
- introduce air and recycled gasses in the lower part of the kiln to flow upward;
- heat the shale to nearly 1,000 degrees F to convert the solid material to liquid;

- collect the upward moving vapor that now contains oil and other products; and
- manage the hot, spent oil shale that continuously leaves the kiln bottom for disposal.

And the Bureau wanted me to model this process to help optimize oil recovery?

Okay. I began by trying to better understand the process and why it worked. The Bureau field engineers were very generous with their time, working with me to better explain the process details. The science behind the oil shale program was explained to me by Andy Decora, director of the laboratory located on the University of Wyoming campus in Laramie.

Finally, the pieces began to fall together. The objective was to recover as much oil as possible while moving the oil through the kiln at a commercial processing rate. Working with the engineers and scientists, we identified three key variables that could be controlled in the process:

1. Size of the oil shale that was controlled by crushing and screening;
2. rate of movement through the kiln that could be controlled by the discharge gates; and
3. quantity and rate of recycled gas introduced near the kiln base.

These were the variables I began to model. Of course, I needed data. The results of experimental runs at the Anvil Points facility were sent to me at my new post in Washington. Over time, I was able to prepare a regression model that reasonably represented the physical kiln in Colorado.

The model proved useful as the Bureau engineers continued to optimize the retorting conditions using the vertical kiln technology. The kiln technology continued to be improved by the Bureau and adopted, and improved, by the industry for commercial use in Brazil. It is still being considered by the industry for locations that need, but do not have, conventional oil deposits.

I moved on to other projects of interest to the Washington staff.

Chapter 3

THE PETROLEUM INDUSTRY OF THE 1960S

OUR TWO HILLBILLY KIDS were in school, and Louise and I were sailing on Lake Washington near Seattle. We were in sailing class, part of my relaxed life as a full-time student at the University. How, you may ask, did this happen?

As it turned out, when I arrived from West Virginia, no one on the Washington staff was interested in learning about the mathematics behind predicting waterflood performance or how to statistically design a research project. These were all professional engineers and scientists, no political appointments. And all had performed as researchers, but all had transitioned to the broader interests important to headquarters management. Rather than perform the work, my boss expected me to learn the skills needed to direct and manage the research being performed at field locations in West Virginia, Oklahoma, Wyoming, Nevada, and California.

I started by understanding the oil shale mining and retorting technology underway at Anvil Points and at Laramie. By 1966, I had learned enough to co-author the paper *Oil Shale: Its Status and Problems,* by J. Wade Watkins (my boss) and Harry R. Johnson. The Bureau at that time was evaluating, with the Atomic Energy Commission, the use of a nuclear explosion to stimulate the production of natural gas from a gas reservoir in New Mexico. A proposed extension of that effort was to explode the bomb under the vast oil shale deposits of Colorado, create a pile of broken oil shale, then retort the shale in place.

Project Gasbuggy was an underground nuclear detonation carried out in New Mexico in 1967 under a natural gas deposit. It was followed by two subsequent nuclear explosions in western Colorado, in 1969 and 1973, in an effort to improve the technology. Field results were disappointing and the gas proved to be radioactively hot. Fortunately, funding for this effort was stopped before the researchers had a chance to contaminate our oil shale deposits.

Automobiles and Air Pollution

Air pollution from automobiles was also a hot topic in the '60s. The Bureau had the only functioning capability in the government designed to understand how photochemical smog formed and how it could be reduced. In fact, the Bureau's Bartlesville laboratory provided early scientific support to the state of California in its efforts to reduce smog in the Los Angeles basin. My boss asked me to learn about this research and how it could be expanded to solve other similar problems.

I was not far up the learning curve when the federal government, led by the Department of Commerce, formed a major study of how automobile pollution could be reduced. Over 100 individuals from throughout government, industry and universities participated in the study. The Departments of

Commerce, Interior, Defense, Transportation, and others all had individuals assigned to the effort. Within the Interior, the official representative was the Director of the Bureau of Mines. Since I could by then speak about photochemical smog, I was selected as the Directors Designee.

The study took nearly a year of intensive study. The findings were released by the Department of Commerce in October 1967, and the detailed subpanel reports in December 1967. Both were titled *The Automobile and Air Pollution: A Program for Progress*. This was my first (but not last) adventure with a major government study. Here you meet really interesting people that can have a major impact on your professional life. In my case, I met Dr. S. William Gouse, Jr., a chemical engineer on loan to the study from the Massachusetts Institute of Technology. We became close friends during the study, and this connection would prove important a few years later as the nation began to respond to oil supply shortfalls.

Higher Education

Meanwhile, I was taking night courses at The American University Center for Technology and Administration. I was guided through the techniques of strategic planning by Dr. Charles M. Mottley, who was on the Bureau Directors staff. It was late in his career, which had included the strategic analyses that led to placing nuclear warheads on submarines as a more flexible alternative than land-based missiles. Working with him and the university, I began to understand there was a broader world to explore than just the narrow focus of petroleum technology. To explore this world, I was allowed to apply to two universities for a year sabbatical from my regular job.

Accepted at both, I could have gone to Stanford for an intensive effort resulting in an MBA. This path seemed like work. Alternatively, I chose the University of Washington, where I only had to audit the class and not take tests. What a deal. I was on full government salary, had a check from the Ford Foundation, and a house in a neighborhood where a lot of Boeing engineers lived. They adopted our family immediately.

I was a rich student and could afford sailing and golf. But the objective was to learn new things that would enhance my future government career. In the engineering school, I learned and practiced the critical skills of discounted cash flow, internal rate of return, and return on capital. In the law school, I learned who owns the Colorado River and the legal issues associated with that ownership. Gladly, I did not need to take tests with those smart young legal minds, but the law professor called on me during every class to make sure I was at least reading the assigned material. I was.

The fellowship program granted by the National Institute of Public affairs was represented at the university by professors from schools of economics, geography, and public policy. We developed an outline of a paper that would tie my knowledge of the petroleum industry with the public policy issues important to the industry, including foreign imports, taxes, and leasing of public lands. I filled in the outline, and Professors Crutchfield, Pealy, and Cooley provided real-time reviews and suggestions. They decided that the paper was worth publication and submitted it to a law journal for consideration. The journal agreed, and *Petroleum in*

Perspective by Harry R. Johnson was published in the Natural Resources Journal, the University of New Mexico School of Law, January 1971.

Chapter 4

STIMULATING OIL SHALE DEVELOPMENT BY LEASING

"Hey, Hey, LBJ, and how many kids did you kill today?" was a favorite chant from the almost daily demonstrators outside the White House protesting the Vietnam War. Larry Burman, my office partner, and I often extended our exercise walk from the Interior Building to see who was protesting, and with what signs and chants. Violent protests were common at university campuses. President Johnson decided not to seek re-election. It was 1969, and the energy world I knew was also about to erupt.

Hydraulic fracturing and waterflooding were introduced in the 1950s. This combination was so successful that the ability to produce oil became far greater than demand. As detailed in my *Petroleum in Perspective* paper, excess productive capacity increased each year for over a decade, from 1950 to 1964, reaching a peak of 37 percent. The U.S. was the world's swing oil producer. In this role, domestic policy limited the volume of oil that could be imported from other countries. Policy also limited the volume of oil that could be produced from domestic fields. Both actions reduced oil supply, which supported the domestic oil price at a level higher than the world price of oil.

So what happened? Domestic oil production peaked in 1970 and began a steady decline. Oil imports had to increase to satisfy growing demand. The cost of increasing oil imports threatened to destabilize the nation's economy. Addressing these concerns, President Nixon in 1971 instructed the Secretary of the Interior to expedite a leasing program that would lead to oil shale development on public lands.

This order came just after Congress passed, in 1969, the National Environment Policy Act (NEPA). This act required an Environmental Impact Statement (EIS) to be drafted and reviewed. The Interior prepared a preliminary oil shale EIS in 1971. The EIS was strongly criticized during hearings on the program. It was found to be inadequate under the law. The preliminary EIS was withdrawn, and the Interior made plans to prepare a more comprehensive document.

The Interior had another major ongoing environmental issue at the time of the 1971 Nixon order to expedite oil shale development. Prudhoe Bay, the nation's largest oilfield, was discovered in 1968, but environmentalists had successfully blocked the approval to construct an 800-mile pipeline to the Port of Valdez. This battle was still raging even as the Interior began to consider how to address oil shale, another project with major environmental impacts.

In the meantime, I was learning a lot about coal technology and its impact on the environment. Working with the current and past directors of the Bureau of Mine coal research program, we began to document the magnitude of the environmental impacts from energy development. We pointed out that by 1965, about 3.2 million acres of land had been disturbed by surface mining; over 80 percent of the thermal pollution of water resulted from fuel use; and nearly 80 percent of air pollution was caused by fuel combustion. Our paper was designed to be a wake-up call to the technical community that when selecting a fuel, the cheapest fuel is not always the best. In addition to costs, the effects of production, processing and utilization must also consider the effects on land, water and air. Our paper, *Fuels Management in an Environmental Age* by G. Alex Mills, Harry Perry, and Harry R. Johnson, was published in the prestigious publication "Environmental Science and Technology" in January 1971.

Prototype Leasing Program Gets Organized

Reid Stone from the U.S. Geological Survey was the Oil Shale Coordinator selected to implement the 1971 Nixon directive to accelerate oil shale development. When the first EIS failed, Reid knew that the mining and retorting technical expertise of the Bureau of Mines engineers would be needed to move the prototype program forward.

The Bureau selected me to mobilize its resources. Thus drafted, I became the Department's Deputy Oil Shale Coordinator in charge of the EIS.

At our first organizational meetings, it was clear that the Interior's lawyers would give the final approval to release whatever document we produced. Our lawyers were then in the third year of a losing fight to get authority to proceed with the 800-mile pipeline in Alaska. A few months earlier, they had been smacked down when the preliminary oil shale EIS was found inadequate under the law. They were in no mood to play nice. They suggested to me that if we were going to rape, pillage and burn the land, say how and quantify the environmental effects.

We had our marching orders. All of the resources of the Department would be needed to contribute to this study:

- U.S. Geologic Survey – Geology and ground water movement
- Bureau of Mines – Mining, retorting, and land, water, and air impacts
- Bureau of Reclamation – Surface water availability
- U.S. Fish and Wildlife Services – Baseline conditions and environmental impacts
- Bureau of Land Management – Leasing terms and regulations

The Department's resources were all readily available, since the Secretary was responsible to the President for producing the EIS. We estimated the EIS would take months to produce and months to respond to public comments. We needed a place to set up shop.

The Bureau of Indian Affairs had a small building just behind the main Interior building that proved to be perfect. We moved into empty office spaces and began to produce an EIS.

Environmental Impacts of Oil Shale Development

The Interior had selected six prototype tracts for leasing before I was drafted, two each in Colorado, Utah, and Wyoming. The Wyoming sites were expected to be developed by in-situ extraction.

In-situ requires no mining. Rather, wells are drilled vertically into the shale formation. Heat is applied to melt the oil shale in place, and the liquid is then pumped to the surface for processing.

The remaining four sites were expected to be developed by mining, followed by retorting. Information needed for the environmental assessment was collected and ultimately published as *Volume III, Specific Impacts of Prototype Oil Shale Development*.

A lead engineer from a Bureau of Mines field facility was selected for each site. Their job was to engineer tract development; road and site preparation; required support surface facilities; how much overburden to remove; size of the mine rooms and pillars; volume of water to pump; and quantity and movement of raw shale to the retorts.

Field specific data were brought to our Bureau of Indian Affairs building. Approaches were presented, discussed, and refined. Once approved, mine development calculations began to be finalized and the work documented. For an underground mine, for example:

> "Entrance to the working area would be gained by 4 vertical concrete-lined shafts 20 to 30 feet in diameter. Each would be about 1,500 feet deep…"

> "Rooms and pillars would be 60 ft. wide." "…full production of 73,700 tons per day would be achieved within the shortest time."

Bureau retort engineers used the mine design to size the oil shale processing facility. For example:

"For a 50,000 barrels/day shale oil production, six retorts each process 505 tons per hour of shale and briquettes and produce 307 barrels per hour of crude shale oil, 4.04 million standard cubic feet per hour of excess low heating value gas (about 100 BTUs per cubic foot), and 410 tons per hour of spent shale."

Air contaminants released during retorting (particulates, nitrogen, sulfur dioxide) was quantified. Water needed to cool the hot shale exiting the retort was calculated. The volume of spent shale was estimated, disposal locations identified on a topographic map, and topography changes identified; for example:

"...canyons B, C, and D would be filled to a depth of about 250 feet..." "...some 10,000 linear feet of State Bridge Draw and 9,200 linear feet of Right Fork of East Fourmile Draw would be needed for the all-surface-disposal option over a 20-year period."

Leo Schrider, my petroleum engineering friend from West Virginia, was tasked with developing the in-situ plan for the two Wyoming tracts. He finished early, and I asked him then to calculate the spent shale holding capacity of specific canyons. This required measuring instruments, blown-up topological maps, and a lot of patience. He did a great job, but was not happy about this assignment.

Given this site-specific information, specialists from throughout the Interior were able, for each of the six tracts, to quantify environmental impacts of the prototype program. For the six prototype tracts:

- The cumulative area affected would be about 6,650 acres;
- Vegetation would be lost: pinyon-juniper 1,400 acres, sagebrush 2,260 acres, and others;
- About 2 tons of airborne particulates, 8 to 12 tons of nitrogen oxides, and 98 to 186 tons of sulfur dioxide would be released each day;
- Wildlife intolerant to human activity (mountain lion, peregrine falcon, and others) would be displaced;
- Annually, 88 to 353 Animal Unit Months of livestock grazing would be lost; and
- The social and economic character of local communities would be changed forever.

If the prototype program was successful, commercial oil shale development was expected to follow. The specific impacts of the six-tract prototype program were then used to estimate the impacts associated with an industry that produces 1 million barrels/day of shale oil.

Volume I, Regional Impacts of Oil Shale Development, quantifies these impacts. Significant impacts include 115,000 new people added to the region, increased urbanization, changes in regional lifestyles, and changes in land use patterns. Just over 700 pages in length, Volume I detailed these impacts and the consequence of industrial development.

Volume II, Energy Alternatives was a critical document, since no alternative in the preliminary 1971 EIS had been prepared. Because of this, the 1971 EIS was found inadequate under the law. The Interior lawyers had a vested interest in making sure that we produced a document that would meet legal requirements.

As the lead author of Volume II, I was able to use a lot of what had already been published. Text, figures and tables from *Fuels Management in an Environmental Age*, *Petroleum in Perspective* and *Changing Investment Patterns of the Petroleum Industry* all found their way into this volume. A full range of technical and policy options were considered.

Rereading Volume II now, I am impressed with our insight on oil imports:

> "…A new dimension to international oil movements must now be considered – the demonstrated ability of the oil-exporting nations to act both separately and in unison to attain specific economic objectives at the expense of both the oil companies and consuming nations. Increasingly, the exporting nations have won price concessions from producing companies which ultimately must be paid for by consumers. The balance has now been tipped in favor of the oil-exporting nations."

Just one year after this was written, in October of 1973, the Organization of Petroleum Exporting Countries (OPEC) imposed a near total ban of oil exports to the United States, causing national panic. Remember the long lines to get gasoline? This act of economic war was a wake-up call to the nation, resulting in major federal actions covered in another chapter of this journey.

Our lawyers almost daily reviewed draft material and offered useful suggestions. Finally, they were satisfied with what we had prepared and recommended to the Secretary that the draft EIS be released for outside review. So in September 1972, the nearly 2,000 page, three-volume draft environmental impact statement was released for public review and comment.

Environmental Impact Statement Gets Reviewed

It seemed like everybody wanted to comment on the draft EIS. Some wanted to improve the draft regulations; all were concerned about environmental impacts; and some wanted to kill the prototype program.

Ninety-five (95) individuals appeared at public hearings held in Colorado, Utah, and Wyoming in October 1972. An official reporter made a verbatim transcript of each hearing. These transcripts totaled 450 pages and were published as *Volume VI of the Final EIS, Public Hearings Held During the Review Process.*

Written comments were received from 205 organizations, including federal agencies, state agencies, environmental groups, industrial companies, and private citizens. These written responses totaled another 1,939 pages and were published as *Volume V of the Final EIS, Letters Received During Review Process.*

The task of organizing all of this review material fell to the most organized person I knew, Dr. Andy Decora, from our Laramie oil shale research center. Andy not only had a science background, but was working on becoming a lawyer. He fit right in with the Department's legal teams.

Andy carefully catalogued each written response and each hearing suggestion. He then created a format to collect and respond to similar comments. Once organized, we were able to respond to comment classes, which often improved our draft EIS. For example, questions were raised about water rights. In response, we added several pages of additional details on the availability of water, the amount each state is entitled to, and the possibilities of augmenting supplies in the upper basin.

Others were concerned about baseline environmental conditions. In response, the Department changed the lease terms to require that additional environment studies be conducted by the lessee for at least 2 years on each tract before development could begin.

Our 200-page analysis of the review comments and the Department's response is contained in *Volume IV of the Final EIS, Consultation and Coordination with Others*.

Reviewing the comments and our response today, I am impressed with the effort made by the Interior professionals to improve the draft EIS material and the prototype lease terms. Our lawyers also approved our efforts and recommended to the Secretary that the final EIS be released.

The final EIS on the prototype oil shale leasing program, now six volumes and nearly 6,000 pages in length, was published August 1973.

Prototype Lease Sales Win Large Bonus Payments

The final EIS was subject to intense review by environment groups opposed to oil shale development. Our sources had them meeting in Denver to see how they could stop the prototype program. According to them, the environmental groups could find no feasible way to halt this program.

The nation was not prepared for the Arab Oil Embargo of October 1973 and demanded actions to increase domestic oil supply. President Nixon signed the Trans-Alaska Pipeline

Authorization Act on November 16, 1973, which granted a federal right-of-way for the long-delayed pipeline. A few weeks later, on November 28, 1973, the Secretary of the Interior released his decision to proceed with oil shale leasing.

The economic war over oil supply was now fully underway.

Prototype lease sales began in January 1974 and were complete by April 1974. Spurred by the Arab Oil Embargo, the industry was determined to win the rights to develop the high resource tracts. Winning bids totaled nearly $450 million. That was a lot of money in 1974, a sum that in today's dollars would be $3 to $4 billion.

This was serious business, and the winning bidders began to prepare detailed development plans. Finally, the environmental community saw an opening to halt development. In December 1977, The Environmental Defense Fund, The Colorado Open Space Council, and Friends of the Earth challenged development of the two Colorado tracts, arguing that additional Environmental Impact Statements should be required. The Federal District Court rejected that argument and stated:

> "We are satisfied that nothing contained in the detailed development plans, as modified or supplemented, demonstrates any significant environmental impacts which were not identified and described in the 1973 EIS."

Chapter 5

UNITED STATES RESPONDS TO THE SAUDI ARABIA OIL EMBARGO

"DARN CHERRY BLOSSOMS," I thought as I wiped them from my windshield in the dark. They were beautiful that morning when I arrived at my assigned parking area near the Tidal Basin, but then it rained. What a soggy mess.

It was 1973 and I had returned from working on oil shale leasing to regular Bureau of Mines duties. We were evaluating the impact of the energy crises on our research programs; for example, the impact of higher prices on coal operations. At that time, the Organization for Economic Cooperation and Development (OECD) decided to reassess the world-wide role of coal, and I was selected to chair one of several ad-hoc working parties. My first trip was to the OECD headquarters in Paris, and it was great.

Then all hell broke loose!

Arab states led by Egypt and Syria launched a surprise attack on October 6, 1973, on Israeli positions on both banks of the Suez Canal, Golan Heights and surrounding regions. Within three days, Israel had mobilized its forces and halted both the Egyptian and Syrian offensive. Supported by massive resupply efforts from the United States, Israeli forces then pushed the invading Arab forces back. The Yom Kippur War was over in a few days, resulting in an Israeli military victory.

Arab nations were not happy with our support of Israel. As punishment, Saudi Arabia and others declared an oil embargo beginning October 16, only ten days after the surprise attack. Oil production was cut back 4 to 5 million barrels per day. A total ban on crude shipments to the U.S. was imposed.

Stating the obvious, President Nixon addressed the nation on November 7, 1973, and declared that we had an energy crisis.

The crisis impact was rapid and intense.

By January 1974, imports to the U.S. fell by 2.7 million bbl/day. Oil prices jumped over 250 percent, from $3 before the embargo to $11 per barrel in January. Without oil, gasoline and diesel shortages rapidly appeared. Motorist waited in long lines for expensive, limited amounts of gasoline. Truckers blocked traffic on major highways, protesting the lack of diesel and its cost. About 500,000 people lost their jobs. Consumer prices skyrocketed. And the Gross National Product dropped $10 to $20 billion during the embargo.

President Nixon proposed Project Independence, with a goal of achieving energy self-sufficiency by 1980. The project was organized at the height of the embargo.

Project Independence

By any standard, Project Independence was a massive government effort. Managed by the Federal Energy Administration, the study ultimately had 500 professionals grouped into 27 working groups and task forces. Bill Gouse chaired the synfuels task force that focused on converting coal to liquid fuels. I chaired the oil shale task force.

It took months to organize this study, which began in early March of 1974. Just as the study got underway, on March 18, 1974, Saudi Arabia announced the end of its five-month oil embargo. Oil supplies to the U.S. began to return to normal.

The Project Independence analysis was the most comprehensive energy analysis ever undertaken by the government. It never presented a blueprint for how the nation could achieve President Nixon's goals. It did, however, provide a framework for developing a national energy policy.

Oil shale commercial development had already been stimulated by leasing of federal lands. Early in 1974, the industry invested $3 to $4 billion for the right to develop these leases. The industry was preparing detailed development plans that would require billions of dollars more for physical development, including the construction of entirely new towns to house new employees. From our oil shale task force perspective, no further government action was needed to achieve a business-as-usual production level of 50,000 barrels per day by 1980, and 250,000 by 1985.

Accelerated development over these target levels would require federal incentives identified in the oil shale task force report. Estimates of the effectiveness of these incentives, their costs, and relative priorities were not required.

The *Project Independence Report* was published in November 1974. Oil was flowing and gasoline waiting lines were a thing of the past. Massive intervention in the energy market would need to wait for the next overseas energy crisis. We only had to wait 4 years.

Energy Research and Development Administration

As Project Independence was being organized, the Department of Interior established an Office of Research and Development to better focus research being conducted by the Bureau of Mines, the Office of Coal Research, and the Geological Survey. Bill Gouse was recruited from Carnegie Mellon to run this new office. Bill remembered me from our earlier Department of Commerce study on *The Automobile and Air Pollution: A Program for Progress*. He asked me to be his assistant director. YES was the answer. I finally had an end office in the Interior building with a window looking toward the State Department! Nothing was better than a parking pass plus a window office for a professional bureaucrat in those days.

On October 11, 1974, President Ford signed the *Energy Reorganization Act of 1974* to establish the Energy Research and Development Administration (ERDA). All six of the Bureau's energy research centers, the Office of Coal Research headed by George Fumich, and other Interior organizations were transferred to the new organization. Here we were merged with the largest energy organization, the Atomic Energy Commission, and one of the smallest, the solar research program from the National Science Foundation headed by Don Beattie.

All told, the ERDA had over 7,000 employees and a fiscal year 1975 budget of $3.6 billion. The entire focus was energy research. Energy policy and its regulation was the responsibility of the Federal Energy Administration, later to be headed by Jack O'Leary, a former director of the Bureau of Mines.

With Bill Gouse, we created a powerful organization to plan and administer all of the fossil energy programs. This 70-person staff focused on strategic planning, engineering assessments, environmental impacts, and budget execution. We were well prepared to support the next government effort to increase oil supplies. It was not long in coming.

Synfuels Fuels Commercialization Program

In his January 1975 state of the union message, President Ford unveiled a program to have the nation's vast coal and oil shale resources contribute to future energy supplies. The goal was to have 1 million barrels of synthetic fuels and shale oil production by 1985, with an incentive program to achieve the goal.

The White House organized this study under its Office of Management and Budget (OMB). This assured the study results would lead to a Congressional proposal for action. One of the bright stars of OMB, Dr. Bill McCormick, was selected to lead the study effort. Bill Gouse was selected to lead two groups. I was selected to lead the shale oil study and was also promoted to lead the effort to integrate study results. In this role, I worked closely with study leader McCormick and other White House staff.

The ERDA and nine other Federal agencies participated in the Synfuels task force that started operations in February 1975. We were able to build on the earlier Project Independence results, requiring only 50 federal employees instead of 500. This time, we were not trying to become energy independent. Rather, we were trying to achieve a specific target production level and to identify what was required to achieve this production.

The *Synthetic Fuels Commercialization Program* report reached the following bottom line conclusion:

> "In the absence of Federally provided economic incentives or other policies creating a stable and favorable investment environment, significant amounts of synthetic fuels are not likely to be produced by 1985."

However,

> "With appropriate incentives, maximum production of synthetic fuels by 1985 could be no more than 1.7 million barrels per day of crude oil equivalent without massive dislocations in the economy."

The incentives were financial plans in which the government helps in capital formation and provides financial protection to industry to design, construct and operate production plants. My engineering economics training from the University of Washington (discounted cash flow, internal rate of return, return on investment, etc.) was being tested. In the final report, each industry sector had an incentive package specific to that sector. For example:

- Shale oil and syncrude: Non-recourse loan guarantee (a loan secured by collateral) and a price guarantee, competitively bid;
- Regulated high BTU gas from coal: up to a fixed percent non-recourse loan guarantee, competitively bid;
- Regulated utility/industrial fuels: Construction grant, competitively bid;
- Unregulated utility/industrial fuels: Non-recourse loan guarantee plus price guarantee, competitively bid; and

- Biomass, unregulated: non-recourse loan guarantee, competitively bid.

The *Synthetic Fuels Commercialization Program* did lay the foundation for massive government intervention in the energy market, but this intervention would need to wait for the next overseas energy crises.

The waiting period for the next Arab oil shortfall had dropped to 3 years.

Chapter 6

CLASH OF GOVERNMENT ENERGY PLANS

IT WAS LATE IN THE AFTERNOON. It had begun to rain and thunder as we climbed the steps at the old post office building. Phil White, Martin Adams and I were on our way to brief Jack O'Leary, who was then head of the Federal Energy Agency (FEA). The thunder outside the building proved to be an omen of the thunder we were about to receive inside the building.

It began earlier that day in the spring of 1977. I was invited to an energy briefing in the West Wing of the White House. Got there late and sat near the top of a long table. As the person to be briefed, Jack O'Leary took the final seat at the head of the table. I knew him from our Bureau of Mines days and we said hello. The briefing was on energy regulation, a subject I knew little about. After the briefing, I suggested that Jack learn more about what the Energy Research and Development Administration (ERDA) was doing to prioritize our $3.6 billion research program. He said get your team together and I'll see you tonight.

As we were developing research priorities, O'Leary was developing a national energy plan at the request of newly-elected President Jimmy Carter. These two independent agencies with Energy in their names drew different conclusions about the cost and future availability of natural gas. Actions taken by each agency impacted the energy markets for years in the future.

Energy Research and Development Administration Plans

Bob Fri became administrator of ERDA in January 1977. He turned to our fossil energy group for ideas about how the agency could establish energy research priorities. Dr. Phil White, my new boss at ERDA; Martin Adams, the head of our strategic planning group; and I looked for ways to meet that request. Martin, a Texas A&M chemical engineer, had a long and successful strategic planning career with ESSO, later to become EXXON. He was perfect to lead the analysis to measure how new technologies under development at ERDA would compete and/or complement technologies already deployed by private technology.

Martin and his two deputies, Bruce Robinson and Dave Beecy, invented a sophisticated analytical approach that was absolutely unique. The approach started with services needed for each market demand sector — residential/commercial, industrial, and transportation — and the price each market is willing to pay for oil, gas, coal, or electricity. Any new industry or ERDA technology will be in competition with conventional ways to satisfy these market demands. This is exactly how the real world market works, and it permitted us to establish the relative priorities of all the ERDA technologies.

We started with a forecast of energy demand and supply using the best available energy computer model at that time. However, no computer model could look 25 years in the future and reasonably reflect how markets, prices, and technology would evolve. But within ERDA, we had experts in each market who could gauge changes in demand and supply over time. The problem was how to organize a human expert model to look objectively into the energy future.

Martin explains in his book *The Chronicles of Martin*, beginning on page 154, how the human model was created and organized for the study to be known as the Market Oriented Program Planning Study (MOPPS). To run the human model, up to 300 ERDA experts and support contractor personnel were dedicated full time for up to two weeks at a time. For the second time in my government career, I needed a separate building to house an analysis effort (the first was use of the Bureau of Indian Affairs building in 1972 to prepare the Prototype Oil Shale Leasing Program).

Early results began to be available in April 1977. As the Executive Director of the study, I began the task of presenting these results internally and, later, outside the agency. We began to see shocking results from the study that would completely change how we viewed energy and the government's role in technology. Many of the ERDA technologies, for example, were targeted to making electricity and justified based on a growth rate for electricity of 5 percent per year. MOPPS results said the electricity growth rate would be much lower, 3 percent per year, and this would change the relative priorities for ERDA-sponsored technologies. Martin and his gang had invented a way to look into the energy future with results that would prove to be accurate over nearly the next three decades. Not quite a Nobel Prize effort, but close.

The key to the analysis was the price and volume of conventional energy sources (oil, gas, coal and electric power) available over the twenty-five year planning horizon. These price-volume estimates provided the threshold against which any new technology had to compete for a share of the future energy market.

Our supply-price forecast for natural gas became the focus of intense interest, both within and outside the government. We had two forecasts for natural gas. Both estimates showed large quantities of natural gas were available at higher prices.

Federal Energy Agency Plans

Jimmy Carter became President of the United States on January 20, 1977. On his twelfth day as President, he signed the *Emergency Natural Gas Act of 1977*. He then named Jack O'Leary as head of the FEA and requested a national energy plan within 90 days.

Facing our most severe winter in decades, natural gas shortages caused thousands of factory and school closings and threatened cutoffs to residential customers. The nation was worried about natural gas supplies and Congress passed, with overwhelming bi-partisan support, the 1977 natural gas act. The Act authorized the President to declare a natural gas emergency to offset anticipated gas shortages and authorized the physical transfer of gas from one pipeline to another, if needed. This was an important Act that reflected the concern of the President and Congress over the future availability of natural gas.

By April 1977, the FEA was well into the final draft of the energy plan requested by the President. This plan was based on shortfalls of natural gas. Therefore, the plan rationed the use of natural gas and authorized its use for only high priority applications, such as home heating. Gas was banned for use in new boilers to make electricity. The plan also encouraged the use of coal to replace gas in existing power plant boilers.

The message: Adequate supplies of gas were not expected to be available in the future. So on that rainy April night, when O'Leary saw our analysis of large quantities of gas available at a price, Martin Adams remembered that he thundered:

"What the hell are you trying to do…"

A heated debate about gas availability followed with no resolution between two independent (at that time) government energy agencies.

Days after our thunderous meeting, President Carter appeared on national TV on April 18, declaring that:

> "The oil and natural gas we rely on for 75 percent of our energy are running out…We need to shift to plentiful coal while taking care to protect the environment…"

Three days later, the President addressed a joint session of Congress and outlined the national energy plan to be submitted to Congress the following week. The *National Energy Program Fact Sheet on the President's Program* featured heavy regulations that would discourage the use of natural gas, and incentives to substitute coal for oil in industrial markets.

I could then understand why Jack was so angry with the MOPPS analysis. It showed unlimited natural gas supplies at higher prices. From his perspective, Jack probably thought we were just a bunch of researchers lacking an understanding of energy markets.

The political heat was about to get ugly.

Natural Gas Politics in the 1970s

We continued with the effort to prioritize the ERDA research. But first, we wanted to recheck the nation's ability to support increased oil and gas production. The U.S Geological Survey hosted a working session at their Reston, VA headquarters. Seventy experts from government and industry were organized to consider potential oil and gas production from fields located both onshore and offshore. The experts gave their best considered judgments about the basin potential and the cost of realizing that potential. Individual results were combined to create an overall supply-price curve. Now we had three curves, all of which pointed to increased gas supplies at higher prices.

Here's where it got ugly!

Our gas supply-price curves were leaked to the Wall Street Journal. The Journal decided to run a full editorial about how, deep in the bowels of the ERDA, researches had found large amounts of natural gas available at a price. The Journal called this ERDA-gate as a comparison to the more famous Watergate stories that toppled President Nixon. After this was published, Bob Fri came to my office and wanted me to get my charts; we were on a day trip to New York. "Why?" I asked. I remember him saying that if the Journal was going to write about the study, they should have the facts. We did our day trip. I don't know if anything changed, but we felt better.

Feeling better was about to change.

Senate Hearings

Many members of Congress did not want to regulate the price of oil and natural gas. The MOPPS gas supply information published by the Journal provided evidence that the government should remove price controls and stop regulating prices. The Senate Energy Committee decided to hold a hearing on MOPPS in June 1977.

Chris Knudsen, the developer of the original, very optimistic gas supply-price curve was asked to testify. Details about this hearing are documented in *The Politics of Mistrust* by Aaron Wildavsky and Ellen Tenenbaum, beginning on page 238.

During this first hearing, Senators Metzenbaum (Ohio) and Durkin (New Hampshire) treated Knudsen as a man of integrity who had been wronged by ERDA bureaucrats seeking only to preserve their own interests. Senator Durkin is reported as saying:

> "Is the arbitrarily and politically motivated rejection of the Knudsen findings, and his summary removal, part of a double cover-up – a cover-up of the true extent of our national gas reserves, and a cover-up by the bureaucrats of that cover-up?"

Needless to say, the Committee questions of Knudsen were gentle.

The second Senate hearing (not covered in the Wildavsky book) was held a few weeks later. The MOPPS management team was sworn in. The Committee questions reflected the belief that we were all nasty bureaucrats and we were engaged in a big cover-up of important data. Of course, a lot of press was present to cover what was expected to be an explosive hearing.

The questions and responses got heated. At one point, feisty Hugh Guthrie, head of our oil and gas research program, told the Senators he was a man of integrity and would no longer take the verbal abuse coming from them. This exchange was duly recorded by the press and played later on local and national TV news programs.

I don't think anything was resolved at this hearing, but I was glad when it was over. Heck, we had already said that gas was available at higher prices. Only the amount available at a particular price was in question.

Briefing President Jimmy Carter

President Carter decided to learn more about the research ERDA was conducting and invited Bob Fri to the White House for a discussion. My boss at the time, Roger LeGassie, was on a mission to Europe and was not available for this meeting. So I got drafted by Bob for support.

We were escorted into the far end of the Cabinet Room, where a projector and screen were waiting. Jimmy Carter entered from the far side of the table, accompanied by Jody Powell, his press secretary. Just the four of us. During the briefing, the President, being a nuclear engineer, asked a number of questions about our nuclear projects. Bob handled these questions easily. He was fluid in responding to questions about conservation, coal, petroleum, and natural gas. I only added some information about oil shale technology. We spent about an hour with the President. I think we convinced the President that we knew what we were doing, leading to his approval of our FY78 budget requests.

I often wonder if I would have been busted if I'd tried to steal the White House coffee cup I had been drinking from.

House of Representative Hearings

The House subcommittee on Fossil and Nuclear Energy Research, Development and Demonstration requested a comprehensive status report on MOPPS and its energy findings. This was an important review, since this is the subcommittee that approves ERDA funding requests. We needed to convince this subcommittee that we knew what we were doing.

The subcommittee met on July 12, 1977, at 9:30 am in room 2325 of the Rayburn House Office Building, Honorable Walter Flowers of Alabama (chairman of the subcommittee) presiding. The witnesses were Phil White, Harry Johnson, and Martin Adams. The entire hearing was recorded and published as a 78-page book titled *Market Oriented Program Planning Study* that is still available from Amazon for about $16.

Phil White introduced the first slide (59 slides were to follow) stating that the purpose of the study "…is to give us a better basis to work out the priorities of ERDA's development and demonstration program and do this in terms of market needs."

I was next introduced to provide the study overview, stating (page 4):

> "All of our analysis starts with a concept called service demand. Here we are illustrating residential/commercial demands that must be fulfilled at any point in time. That is, as we go from 1975 to the year 2000, the growth rate for residential units is somewhere around two percent per

> year and the growth rate for commercial is the same. What this does is fixing at any point in time the number of square feet that you have to heat or the amount of units that you have to build."

This concept of service demand was further clarified in response to a subcommittee member question about the rate of growth for electricity (page 20):

> "Mr. Harkin: It seems to me the utilities are saying (the rate of growth) is something like 5 percent, if I am not mistaken.
>
> Mr. Johnson: Yes, sir. We understand that. What we do now is start with service demands, compute clear back to primary demands... We don't put a number in at the beginning and say it is going to be 5 percent. We say we have got to provide so many services... it will require 3 percent per year central station electricity... It is a very controversial finding, particularly within an agency which is heavily oriented towards producing electricity."

The role of energy conservation was explored in depth during the study. It can arrest demand growth, but if these conservation technologies do not come in, then we are going to see their impact reflected particularly on increased demand for oil and natural gas (page 20).

Industry was by far the largest user of energy by 2000. Its requirements of 48.7 quads by 2000 were almost equal to the energy demands of the transportation and residential/commercial combined.

A quad is a unit of energy equal to one quadrillion BTU and is used as a standard method for comparing energy forms. For example, one quad of coal is equal to 36,000,000 tons, while one quad of diesel oil is equal to 5,996,000,000 gallons.

For the industrial sector, with conservation, energy demand drops from 48.7 quads to 35.8 quads, a 26 percent savings of 12.9 quads by 2000, with significant changes discussed on page 7:

> "There are a number of changes that take place in this sector…we see a big reduction just due to process changes, but most interesting is the category called waste heat and material, which becomes significantly more important over time as industry begins to take waste gas…and to recycle them to generate either heat or electricity, the so-called cogeneration.
>
> Additionally (on page 17) we see the introduction of atmospheric fluidized boilers coming into this market, a significant impact by 2000, waste materials being used, biomass, geothermal steam, nuclear, liquidized feedstocks from coal, pressurized fluidized bed, and conventional coal.
>
> We see, surprisingly, not a large reduction in oil…We have discovered that a lot of the oil required and part of the gas is for feedstocks. ERDA has a large coal liquefaction program, as you are well aware, and what we are trying to do now is to understand what kind of molecules we need here, what kind of chemicals we actually need…we are trying to look at perhaps the coal feedstock program to satisfy the demands in the industrial sector."

Residential/commercial was expected to require 25.3 quads of energy by 2000. Significant conservation savings are discussed on page 17:

> "...we saw a savings of 5 to 7 quads...it is composed of various conservation packages. Conservation I are the things that you can do easily, like home insulation. Conservation II is a little bit more difficult conservation technology to put in. Conservation III are the ultimate...conservation packages...automatic timers to move heat and air through commercial buildings, load management in buildings, very sophisticated packages of technologies. Each of these has a specific technology in back of it. The list runs, I guess, 100 to 150 individual technologies that we group together for ease of handling the data, but we do have all the data that backs it up."

Transportation was expected to require 27.7 quads of energy by 2000. Significant transportation changes were discussed on page 17:

> ...we had about a 7 quad savings in this sector with most of it attributable to the heat engine propulsion, but there are operating, air engine and air design changes, chassis body changes, things that will make our vehicles and our airplanes more streamlined; all are expected to make impacts.
>
> ...other (transportation) conservation things are in pipelines, new motors, maybe friction-reducing additives. There in some conservation from railroads contained in this category."

Demand for baseload electricity and the baseload technologies was a specific focus of the study and, as reviewed on page 17:

> ...nuclear, conventional light water reactor, conventional coal, conventional geothermal, and hydro all are the primary contributors to the base load electricity generating sector. We don't see the significant penetration of almost any new technology in this sector, which is certainly important for ERDA which has a large part of its energy budget devoted to the generation of electricity.
>
> Mr. Lujan: You are not showing any solar at all.
>
> Mr. Johnson: We are not showing any significant amounts of solar in base load electricity, that is right."

Oil became a focus of committee questions and, on page 21:

> Mr. Johnson: Oil as we look at it is our major continuing problem. We have to import oil under all of our cases. The question is how much? We don't see significant production increases in domestic resources..."
>
> Mr. Watkins: Even at higher prices; how can you substantiate that?
>
> Mr. Johnson: Because we are depleting our resource in oil, and we don't see the resource available out there to be called upon, even at higher prices.... It is as simple as that, and we are in trouble with oil in this country, and regardless of how much we wish it was out there, we just don't see it being there."

Natural gas had far greater prospects (page 14):

> "...In contrast to conventional oil, we do have a number of alternatives... The Devonian shale might be about 50 Trillion Cubic Feet (Tcf), again a large resource that underlies most of the East Coast....this is an area that we are exploring and delineating the resource right now, and again this uses conventional technology.
>
> The second commodity is methane from coal seams....You get it by degasifying before coal mining. It is fairly inexpensive....The gas is contained both in the East and in the West. It again is a resource that can be tapped, largely using conventional technology, although we might have to drill deviated holes, and use fracturing.
>
> The third alternative is Western tight sands...There are 11 basins in the East and the Midwest, all of which have their own particular problems...We are still looking for a technology."

Gas deregulation became a focus of the subcommittee interest (page 23):

> Mr. Flowers: One thing that is running through here that you all have carefully hidden is deregulation and its impact. Let's talk about that a little bit. What would deregulation do to supplies? What is the bottom line on that?
>
> Mr. Johnson: Our objective is not to resolve the deregulation-regulation question. It is to prioritize our research.
>
> Mr. Flowers: What if you had in 1985 $3 per thousand cubic feet?

> Mr. Johnson: $3? Our analysis would indicate we would have more gas available than anybody could use at that time.
>
> Mr. Wirth: Would you say that again?
>
> Mr. Johnson: Our analysis would indicate we would have more gas available than anybody could use at that time.
>
> Mr. Lujan: Natural gas?
>
> Mr. Johnson: Natural gas.
>
> Mr. Wirth: This is in 1985?
>
> Mr. Johnson: Yes. Now there is a timing development. It takes time to get this."
>
> Mr. Flowers (page 26): I think it is absolutely ridiculous to be burning (unregulated) $15 a barrel Arabian oil to make electricity in Massachusetts when you can be burning $3 a thousand cubic feet Oklahoma gas.
>
> Mr. Lujan: Three cheers for Oklahoma."

Concluding remarks (page 32):

> Mr. Flowers: Gentlemen, let me say, obviously it is an interesting subject and I hope you will develop it further because I think this kind of planning is the key to the whole picture. I have enjoyed this morning. I wish we had a better turnout of members. I do not think they were aware how stimulating the discussion would be.

The waiting period for the next Arab oil shortfall had now dropped to one year.

Chapter 7

DEPARTMENT OF ENERGY CREATED

LIKE MOSES COMING down the hill with the Ten Commandments, Administrator Bob Fri returned from a mountaintop retreat with his version of Energy Research and Development research priorities. His analysis set off a chain of actions resulting in the FY 1978 budget request. And this budget, in turn, led the nation toward energy independence using technologies that include horizontal drilling to liberate oil and natural gas from our vast unconventional shale deposits.

Strategy for Energy Technologies

It was quite a review, all hand drafted with a working date of 8/22/77. Using a detailed briefing book prepared by Martin Adams and the Market Oriented Program Planning (MOPPS) team, Bob Fri prepared a strategic background to guide ERDA research priorities:

> Liquid Strategy – Nearer term, Enhanced Oil Recovery, residential/commercial, and industrial conservation are essential and about the only things that will help. Longer term, strategy must be to access higher priced synthetic liquid. A policy of high price, phased in over time, is

unpleasant but essential to accelerate conservation, especially in the Northeast and take advantage of potentially large gas supply. Shale should be the first priority option, being both cheaper and amenable to production of transportation fuels for improved heat engines. Coal liquids and biomass are secondary but important hedges, but must be directed toward transportation fuels and feedstocks.

Coal Strategy – The constraint of coal use is twofold: environment and the "hassle factor". Principal targets are for industry and peak/intermediated electric. Some priority should be given to less sophisticated but still important clean coal technologies (atmospheric fluid bed).

Gas Strategy – All things being equal, there is some time to develop high BTU gas and relatively little need for medium and low BTU gases. Unconventional gas (Devonian Shale) needs characterization of price and quantity to determine if high BTU gas from coal is needed at all. However, several factors mitigate in favor of developing some new gas technologies fairly rapidly, including a perception that gas is unavailable, which is wrong.

Electric Strategy – Coal and nuclear dominate this sector, with regionally important geothermal and hydro showing strength. Storage appears important for peak/intermediate based leveling. Essential strategy must be to preserve coal and nuclear options, in view of coal's environmental and nuclear uranium supply constraints. Short term, this boils down to environmentally acceptable coal, which equates to flue gas desulfurization and waste disposal for nuclear. Longer-term, 1990 development of other options is needed as a hedge. For coal, clean coal tech:. nuclear, an advanced reactor.

<u>Conservation Strategy</u> – Of central importance are improved heat engines, industry conservation, non-R&D residential/commercial conservation, and removal of institutional barriers. Beyond this, only a few major technology thrusts seem important.

Bob Fri then prepared a detailed budget based on his strategic analysis. Changes from the actual FY 1977 budget to the FY 1978 budget request reflected his understanding of what ERDA could do to support future energy markets. He allocated large budget increases to conservation and fossil.

We had less than a month to integrate the ERDA and FEA budgets and present a preliminary draft to the President's Office of Management and Budget (OMB). My notes show I met with Jack O'Leary and Bob Fri to resolve major issues. This was followed by a meeting with Dr. Schlesinger, the new Secretary of the Department of Energy, to resolve remaining issues. One key issue was the size and fill rate for the Strategic Petroleum Reserve then being created along the Gulf of Mexico.

We were able to get an FY1978 budget to the OMB in a timely manner. However, the budget reflected a merger of two different organizations with different budget needs. Negotiations with the OMB to resolve issues continued up to late December, when the budget documents needed to go to the printer.

In the end, we had created and defended a research budget that was consistent with MOPPS findings. Congress ultimately agreed and passed the DOE budget with large increases in the conservation and fossil research programs.

DOE Created, But What About George?

On August 4, 1977, President Carter abolished the Federal Energy Administration and the Energy Research and Development Administration. James Schlesinger was sworn in as first Secretary of Energy the following day.

Planning for a new DOE got underway immediately with the nomination of names to fill key positions. Phil White was nominated as the first Assistant Secretary for Fossil Energy and I was nominated to be his Deputy Assistant Secretary. These names went forward for Senate approval.

But what about George, Senator Byrd from West Virginia asked as he reviewed the nominees? The question was answered on September 12, 1977, when President Carter nominated George Fumich, also from WV, to be the Fossil Energy Assistant Secretary. George was an expert in coal technology and my friend and colleague starting with our Bureau of Mines days. He began to organize Fossil Energy while I continued to help integrate the DOE and ERDA budgets.

Project Implementation Moves to Field Locations

I had, by this time in my journey toward energy independence, a comfortable window office looking at the Smithsonian Castle, a parking space inside the Forestall Building (no more cherry blossom issues), and the highest possible government pay grade level. Not bad for a professional bureaucrat who had never been asked party affiliation.

Two events were about to change that comfortable feeling.

First, the Civil Service Reform Act of 1978 created the Senior Executive Service (SES). According to the Office of Personnel Management, SES members serve in key positions just below the top Presidential appointees and are the major link to the federal workforce. That was great, but I liked being a GS-18. Didn't matter; I was drafted into the first SES class.

Second, George Fumich had finished his reorganization of the Fossil Energy staff. George knew a lot about coal, but little about the oil and gas technologies that were about to get a major increase in funding. He asked me to move to Oklahoma to manage this new program.

Before I accepted, I had to convince my wife Louise that all of Oklahoma was not flat. We drove from Tulsa 30 miles north to Bartlesville, and I pointed out some hills and some trees. Convinced that we could survive Indian Territory, we decided to move from our comfortable home in Maryland to our new home in Bartlesville. And that's how, in 1978, I became the Director of the government's central oil and gas facility.

The next Arab oil shortfall was now only months away.

Chapter 8

UNITED STATES RESPONDS TO IRAN CIVIL WAR

THOUSANDS HAD GATHERED near the parliament building to demand free elections when the first shot was fired. More shots followed. The crowd dispersed, leaving behind several hundred dead bodies. This day, September 8, 1978, became known as Black Friday and marked the point of no return for the revolution in Iran.

Widespread demonstrations and rioting followed. Rioters took over the oilfields and, by December 1978, all oil exports from Iran were cut off. In the month following, January 1979, the Shah of Iran abdicated his throne and fled the country.

The United States was not prepared for this loss of oil, just five years after the 1973 Yom Kippur War embargo.

Oil production from the lower 48 states (both onshore and offshore) continued its long decline from its 1970 peak production. From the Saudi oil embargo in 1973 to the Iranian civil war in 1978:

- Domestic oil production declined 0.5 million barrels per day (a 1 million barrels per day gain from Alaska was

more than offset by the continuing decline from the rest of the U.S. oilfields),
- U.S. oil production could not keep up with rising demand. The difference between demand and supply had to be made up by oil imports, which increased 2 million barrels per day, and
- Oil imports climbed from 34.8 percent of consumption to 42.5 percent. At the time of the 1978 revolution, the U.S. was importing 800,000 barrels per day from Iran, or about 10 percent of our total net imports.

Without oil from Iran, the industry began to draw down oil stocks stored at refineries and at central storage located in Cushing, Oklahoma. However, as these stocks became depleted, the oil needed to supply the nation's refiners began to run out. Shortages of gasoline and diesel appeared across the nation. Oil, gasoline, and diesel prices doubled overnight. Long lines for gasoline returned.

Facing diesel shortages as high as 45 percent, truckers began a strike on June 21, 1979. Meg Jacobs describes what followed in her book *Panic at the Pump*, beginning on page 216:

> "The strike was already escalating. Police reported shootings, destruction of property, and intimidation in dozens of states. Commerce slowed across the country, with as many as seventy-five thousand truckers refusing to drive. Transportation in California came to a standstill as fruits and vegetables perished on the vine. The Midwest livestock markets shut down.
>
> On June 23, a gas riot erupted in Levittown, Pennsylvania, a working class suburb of Philadelphia, when twenty truckers blockaded an intersection, flanked

by four gas stations, to protest the fuel shortage. One striker, with American flags adorning his truck, stood on the hood, egging the protesters on. Local residents joined the truckers, chanting, "More gas! More gas!" They threw rocks, beer bottles, and cans at the local police. The crowd grew to two thousand, and police arrested 69 people for disorderly conduct. On the second night, the riot turned more violent, resulting in the nonfatal shooting of an eighteen-year-old. The crowd set two cars on fire and vandalized the local gas stations.

There is a panic at the pumps…"

The United States had to act, and it did.

Massive Synfuels Program

Step one was a massive government program to stimulate the commercial development of synfuels.

The technical and cost basis of the program was established four years earlier with the publication of the *Synthetic Fuels Commercialization Program* report. All of the financial incentives discussed in that report (summarized in chapter 5 of this journey) were available, if Congress would authorize and fund the effort.

Senator William Proxmire introduced the Energy Security Act on April 9, 1979. It passed the Senate on June 20 and the House on June 26. The Act was signed into law by President Jimmy Carter on June 30, less than 90 days after its introduction.

Shows what can happen in a crisis atmosphere!

The crisis began to disappear the following month with increased production from Saudi Arabia. By the end of July, lines at the pump evaporated. However, the nation had already put in place programs expected to lead to increased oil production.

The Energy Security Act authorized $88 billion over twelve years, with $20 billion available in the first five years to support private development of synthetic fuels from coal, oil shale, and tar. This was an enormous amount of money in 1980, and even more in 2016 dollars. With equal cost-sharing by private industry, the first authorized $20 billion would total $230 billion in 2016 dollars. The next $68 billion grows to $560 billion with industry cost-sharing.

The nation had authorized a massive synfuels program that might have cost the government and industry nearly $0.8 trillion in today's dollars. Now that's a real infrastructure stimulus program!

While the Synfuels Corporation was being organized, the Department of Energy began to solicit bids to build synfuels plants. Two oil shale projects were selected for loan and price support in 1981, and one coal gasification plant was selected for a loan guarantee in 1982. When it became operational, the Synfuels Corporation assumed responsibility for these three plants and added three additional projects: one coal gasification plant in 1983, one coal gasification plant in 1984, and one heavy oil plant in 1985.

Of the six synfuels plants to receive loan and/or price guarantees in the 1980s, only one synfuels plant would survive beyond the year 2000.

Enhanced Oil Recovery for Conventional Oil

Step two was government programs to stimulate oil production from conventional reservoirs using Enhanced Oil Recovery (EOR) techniques.

Under the 1977 National Energy Plan, incremental EOR recovery from old fields was free of price controls. The industry increased EOR applications, and the incremental production was sold at unregulated world oil prices that were higher than the regulated domestic oil price. The industry used this incentive program to help build, at a cost of over $1 billion, a 2,200 mile CO_2 transmission and distribution pipeline infrastructure to link natural sources of CO_2 in New Mexico with EOR projects located in West Texas. This program was administered by the DOE.

When I arrived at the Bartlesville Energy Technology Center in October 1978, the DOE budget had already been approved by Congress. Funds available for EOR and Western tight gas sands research (in 2016 dollars) had increased 400 percent, from $40 million in 1977 to nearly $200 million in 1978. Funding continued at a high level, and we were able to start oil and gas projects cost-shared with industry partners. Government funds totaled $0.8 billion while the industry contributed $1.4 billion, for a total program cost of $2.2 billion.

The Windfall Profit Tax Act of 1980 provided tax incentives to encourage the use of EOR. Since this program was a tax incentive, it was administered by the Internal Revenue Service. To support this program, our petroleum engineers prepared definitions of what would qualify as an EOR project. The industry was able to self-certify projects against these definitions and start field projects.

This tax incentive program was wildly successful. The industry applied low-risk EOR approaches (steam recovery) and dusted off high-risk approaches (chemicals) to increase oil production. In the end, four hundred nineteen (419) qualified EOR projects (summarized in the 1986 report *Enhanced Oil Recovery* by T.M. Doscher and J.A. Kostura) were started, at a total cost of $18 billion in 2016 dollars.

The lower-cost EOR projects would survive and are still delivering oil in 2016.

Oil Recovery Engineering

Primary production follows discovery where the natural gas in the reservoir expands and forces oil to the surface for production. Free-flowing oil production is followed by pumping. However, at some point, the natural gas remaining in the reservoir becomes depleted and is not sufficient to force the oil to the wellbore for production.

Secondary recovery represents a second crop of oil. It requires the injection of water or steam to force some of the remaining oil toward a production well. Oil mobilized in the reservoir is initially produced without water. Over time, water breaks through and begins to be produced in increasing quantities. Most waterfloods today produce more water than oil — the oil and water are separated, and the water is re-injected.

Tertiary recovery is a third crop of oil. A chemical is injected into an ongoing waterflood and the chemical mobilizes some of the remaining oil, causing an increase in oil production. These techniques are generally classified as Enhanced Oil Recovery.

Recovery efficiencies for various approaches are documented in *Outlook for Enhanced Oil Recovery* by Harry R. Johnson.

Water and steam injection are the least costly and are highly predictable. Since the oil production response is predictable, project economics can easily be determined and an investment decision made. An EOR chemical like CO_2 is relatively inexpensive if it is produced from a natural source. Other EOR chemicals are manufactured from oil and are, therefore, expensive to produce and use. Moreover, the ability to predict oil production response was low when I arrived in Bartlesville.

We set out to build a database that contained the technical information needed to model the performance of EOR processes. From our Washington staff, I recruited Bob Folstein, who had a degree in chemical engineering and a post-graduate study in nuclear engineering. Bob had spent 25 years working on Central Intelligence Agency projects for which he was cited by then-director George W. Bush as doing a good job, but the Director would not say what he did. Bob was the key to helping organize the Center to support our expanding EOR program.

Don Ward, a petroleum engineer, had spent his early career at the Bartlesville Center and was happy to return from Washington to Oklahoma to head up extraction research, including our cost-shared EOR program. Barbara Barnett joined me from Washington to handle personal issues associated with the reorganization of the Center. I was about to be legally responsible for the management of nearly $1 billion of government funds, and I hired Ron Olson for the legal advice I needed to keep me out of trouble.

Together, we set out to manage a large government program with the internal personnel twists and turns detailed in Chapter 8 of the book *Bartlesville Energy Center, The Federal Government in Petroleum Research 1918 – 1983.*

Our EOR database grew rapidly with high-quality technical data from 26 cost-shared EOR projects with industry partners. We also received quality information on 63 EOR projects conducted in Venezuela under an agreement negotiated in Bartlesville, with our lead negotiator being my friend and MOPPS partner, Martin Adams. The agreement served for years as a model of international cooperation that could have been extended to Mexico and Canada.

Understanding the mechanisms of chemicals displacing oil was being attacked by a large and diverse university research program. We placed focused research contracts with Penn State, Pittsburgh, and West Virginia on the East coast, Stanford and Southern California on the West coast, and Oklahoma, Oklahoma State, Tulsa, Texas, Texas A&M and others in the Midwest.

A mountain of technical information was becoming available, and the information was being incorporated into our growing database. Predictive models were being improved using the high-quality field tests. University results improved the modeling of how chemicals, oil, and water mix in the reservoir. Now we needed to make this information available for use by the petroleum industry.

As part of the cost-shared contract, producers were required to present current field results at an annual meeting supported by the Department of Energy (DOE). The Society of Petroleum Engineers (SPE) held a smaller annual conference focused on research. The SPE suggested that the conferences be combined, which made good sense.

I co-chaired, with SPE President T. Don Stacy, the first SPE/DOE joint conference held in Tulsa, OK in April 1982. Over 2,300 professionals attended the first conference and were treated to a wide range of technical options to consider for increasing oil production. Paper topics included:

- Polymer Flood Project Design in Oklahoma,
- Surfactant Flood Pilot Test in Illinois,
- Polymer-Augmented Alkaline Flood in Wyoming,
- Micellar-Polymer Project in Oklahoma,
- Miscible CO_2 Displacement in Louisiana,
- CO_2 injection for Tertiary Oil Recovery in West Virginia,
- Monitoring Method for CO_2 Floods in North Dakota,
- Successful In-Situ Combustion Project in Louisiana,
- In-situ Combustion in Diatomite's in California,
- Surface and Downhole Steam Generating Techniques Evaluation,
- Performance of the M-6 Steam-Drive Project in Venezuela, and
- Enhanced Oil Recovery From Offshore Gulf of Mexico Reservoirs.

The projects covered a wide range of methods to enhance oil recovery from reservoirs throughout the U.S., from east to west (West Virginia to California) and north to south (North Dakota to Louisiana, onshore and offshore reservoirs). One feature was the status of a large steam drive project presented under our cooperative agreement with Venezuela. Bartlesville technology specialists distributed 8,500 technical reports at this conference. The joint SPE/DOE conference is still being held annually in Tulsa, OK.

In addition to the annual conference, the Bartlesville Center averaged over 7,000 professional visitors each year and distributed over 140,000 reports each year (see *Liquid Fossil Fuel Technology* quarterly report). Our technology transfer program was well received, and domestic oil production was increasing. The high visibility gave me, as the Center Director, a listing in Who's Who in American Science in 1982 and Who's Who in America in 1983.

Things were humming and this lasted till mid-1980, when the EOR got hit with the realities of the marketplace.

Arab Response to the U.S. Actions

Shortfalls of oil into the economy forced the United States to act, and it did with the:

- *National Energy Plan* that significantly changed the energy markets through regulation. Demand for liquid fossil fuels fell 6 percent between 1975 and 1985; demand for natural gas fell 11 percent; while coal demand increased 38 percent as coal replaced oil and gas to generate electricity. Industry applied aggressive conservation measures that cut industrial sector energy demand by 5 percent. These regulations effectively shifted how the nation used fossil fuels while holding overall energy demand to only a 5 percent increase over the ten-year period 1975 to 1985,
- *Incremental Oil Production* under the National Energy Plan was free of price controls. The industry used this incentive to help build, at a cost of over $1 billion, a transmission and distribution infrastructure to link natural

sources of CO_2 with enhanced oil recovery projects in West Texas,
- *Enhanced Oil Recovery Research* funding by the Department of Energy continued at a high level and 26 projects cost-shared with industry partners were started, having a total program cost of $2.2 billion in 2016 dollars,
- *Energy Security Act* authorized the Department of Energy to initiate a massive government effort to stimulate the commercial development of synfuels. Six synfuels projects were selected for development under a program that might have cost the government and industry $0.8 trillion in 2016 dollars,
- *Windfall Profit Tax Act* provided tax incentives to encourage the use of EOR. The industry started 419 projects across the nation at a total cost of $18 billion in 2016 dollars; and
- *Decontrol of Crude Oil and Refined Petroleum Products* on January 28, 1981, by newly elected President Reagan, started a U.S. drilling boom of unprecedented proportions. Rigs drilling for oil and gas doubled from about 2,200 in 1978 to an all-time high of 4,500 by the end of 1981.

The U.S had responded to the energy threats to our economy with regulation, RD&D funding, tax incentives, and executive orders. These actions began to increase U.S. oil production from its low in 1976. Over the nine-year period through 1985, oil production in the U.S. gained 10.3 percent, or +840,000 barrels per day. EOR production contributed about 400,000 barrels/day of the increased production. It looked like we were on our way toward energy independence.

Fig 1. Government Incentives Increase U.S. Oil Production

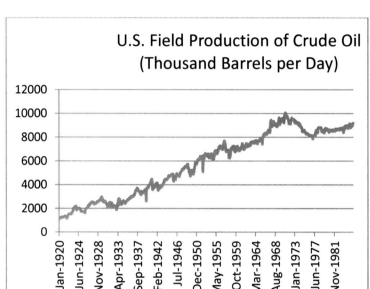

But the Organization of Petroleum Exporting Countries (OPEC) had other plans. They realized the U.S. had turned from paper studies following the 1973 Arab embargo, to actions and funds that would actually increase oil production. They made a decision to protect their share of the world oil market by driving marginal producers out of business.

OPEC began to flood the world market with oil. Oil prices peaked in 1981 at a real price reported by the Energy Information Administration of $61 per barrel. Prices fell 15 percent by 1982.

On May 2, 1982, Exxon announced it was closing its Colony Oil Shale project, laying off more than 2,200 workers. Including lease bonus payments from the 1973 Prototype Oil Shale Leasing Program reviewed in chapter 4 of this journey, the estimated project cost was over $6 billion in 2016 dollars. No commercial shale oil had been produced. Four more synfuels plants would terminate operations by 1998; only one would survive.

By July 1982, Penn Square Bank of Oklahoma failed and the ripple effect caused 139 other banks in Oklahoma to fail.

President Reagan ran on a campaign pledge to dismantle the Department of Energy. While he did not succeed, he did drastically cut the fossil energy budget. Congress restored funding to keep the DOE centers open in 1982-83, but the Department was forced to transfer the Bartlesville, Laramie, and Grand Forks government facilities to private operations. Having worked to at least keep the lights on at Bartlesville, I resigned in 1982 to move into private industry consulting and domestic oil production.

Oil markets were still oversupplied in 1982 and oil prices continued to fall.

By October 1983, the First National Bank of Midland Texas collapsed, making it the second-largest commercial bank failure in American history.

Late in 1983, governors of oil-producing states requested that DOE help to determine state tax incentives that would help keep oil fields operating, even at low oil prices. My successor at the Bartlesville Center, Tom Wesson, led this effort in collaboration with Tim Dowd, Executive Director of the Interstate Oil Compact Commission.

Analysis of the costs and benefits of various incentive options required over a year of intensive work, led by my friend Jerry Brashear of Lewin and Associates (later ICF Resources) and supported by Khosrow Biglarbigi, Alan Becker, and Peter

Crawford. Based on this work, eleven (11) states implemented oil recovery incentive programs to coincide with federal incentives already in place (see *Incentives, Technology, and EOR at Lower Oil Prices* by J.P. Brashear, A. Becker, K. Biglarbigi and R. M. Ray, published in the J. of Petroleum Technology, Feb. 1989). The incentives helped, but oil production continued to fall.

In April 1986, President Reagan abolished the Synthetic Fuels Corporation. Over 40 years, the costs of various efforts to create synthetic fuels may have totaled over $50 billion in 2016 dollars.

In the winter of 1986, oil prices collapsed again, falling from $24 to $12 per barrel. This was the final straw in my efforts to produce oil from a 900-acre Kansas property I was developing by waterflooding. Like thousands of other independent operators, I closed this operation under the bankruptcy laws, but eventually repaid the entirety of the bank loans that had been taken to support field development.

U.S. oil production resumed its natural decline that had started in 1970:

Fig. 2. Cheap Foreign Oil Forces U.S. Oil Production Lower

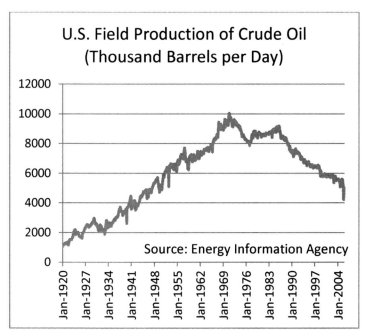

The OPEC strategy to flood the world with cheap oil worked. High-cost chemical EOR projects were terminated. Exploration for conventional oil was greatly scaled back. Oil from Alaska peaked and began to decline. Massive government intervention in the energy market was able to arrest the natural oil production decline, but only for a few years.

The Market Oriented Program Planning (MOPPS) insights about oil proved to be correct. As described in the final 1977 MOPPS report:

"Domestic oil production will continue to decline. America will become more dependent on imports, growing to more than 50% of our oil consumption over the next decade. Enhanced oil production technologies (special liquid or gaseous injection, fracturing, offset and multi-plane horizontal drilling technologies onshore and offshore, and deeper drilling) will help mitigate the level of imports required, and should be considered critical technologies."

Lasting Memories of this Volatile Period

One lasting legacy of the government EOR program is the continuing growth of CO_2 to recover light oil. Industry had established the technical feasibility of this approach in the 1970's in the Permian Basin of West Texas. However, large volumes of CO_2 were not available, and this limited oil recovery to a steady production rate of about 28,000 barrels/day by the 1980s. Large volumes of CO_2 began to be available with the construction of the CO_2 pipeline infrastructure partly funded under the National Energy Plan. Oil production by CO_2 injection continued a steady rise from the 1980s, reaching 280,000 barrels/day in 2011 and accounting for 6 percent of U.S. production.

The Dakota Gasification Plant located in North Dakota, was started under the synfuels program and continues to use lignite coal to produce natural gas sold into pipelines supplying the east Coast. However, as a byproduct, the plant produces large quantities of CO_2 that is transported to and injected into an oilfield just across the U.S. border in Saskatchewan, Canada. The byproduct gas increases revenues by about $30 million per year, while the CO_2 is being used to recover total incremental EOR oil of 130 million barrels.

I first met Khosrow Biglarbigi when he walked into my Bartlesville Office after I had left the government. We were trying to solve a difficult technical issue associated with our waterflood. He took our information, went off for a few hours, and came back with a creditable solution. What I did not know at the time was that he had a photographic memory and was, in effect, a walking petroleum engineering textbook. He used stored information to solve our problem. This was the start of a professional and personable relationship that has now lasted over three decades.

Our first major consulting project was a comprehensive analysis of the Sho-Vel-Tum Oilfield of Oklahoma, supported by the Interstate Oil & Gas Compact Commission and the Department of Energy. Truly a giant among oil fields in the lower 48 states, the field had produced 1.24 billion barrels of oil, and advancing EOR technology was expected to assure its prominence as a source of domestic oil well into the twenty-first century (see *Primary and Secondary Recovery in the Sho-Vel-Tum Oilfield, Oklahoma* by Harry R. Johnson, Khosrow Biglarbigi, Loren Schmidt, R. Mike Ray, and Steven C. Kyser). The analysis was judged to be important by the editors of the Oil & Gas Journal, who asked permission to summarize it for their magazine. Their summary appeared in the Journal in two parts: *Profile of a giant: rising again* on Dec 15, 1986, and *Reservoir/fluid characteristics favor enormous long-term recovery potential* on Jan. 19, 1987. Today, Sho-Vel-Tum is under active re-development, led by Harold Hamm and his Continental Resources company, with horizontal well drilling technology.

Drilling and completing a vertical well to oil and gas trapped below the surface is an example of the conventional approaches practiced since Drake drilled his first well in 1859. Conventional oil resources have provided, and continue to provide, a dependable flow of oil needed to support the nation's economy.

But it is our vast unconventional oil deposits that are now leading the nation toward real energy independence, as next described in this Journey Toward Energy Independence.

Chapter 9

UNCONVENTIONAL NATURAL GAS PRODUCTION

LITTLE NELL, a petite female fish, was floating on a vast inland sea, soaking in the last rays of the setting sun. The sun's light energy had already taken carbon dioxide + water and converted them to sugar + oxygen when Little Nell died and she sank slowly to the sea floor. Her little body had absorbed the sugar, and the sugar had a lot of carbon. Given the right temperature and pressure conditions, the carbon could be converted to oil or natural gas. Little Nell had just started a 400 million year journey from the Devonian Age to modern times.

Deposited, Mashed, Heated, and Fractured

On the sea bottom, Little Nell began to decompose and mix with clay and silt. She was joined by billions and trillions of other plants and animals and, together, they slowly began to build an oil and gas source. This mass may have increased to several thousand feet in thickness as the nearby mountain range began to erode. Rivers fed by rain water carried sand, silt, clay and other eroded materials that covered and compressed Nell and her friends, forcing them deeper toward the hot center of the earth.

Little Nell, being on the bottom, was cooked into gas, while others near the top of the pile were cooked into oil. The pile, mashed to about 100 feet in thickness, is now known as the Marcellus Shale.

Then, only ten thousand years ago, ice probably a mile or more in thickness forced Nell and her friends deeper into the earth. As the ice advanced, it gouged out early drainage systems, creating the Great Lakes. Eventually, the ice began to melt and retreat toward the Arctic.

In the Appalachian Basin, the earth rebounded from the intense pressure of the retreating ice. The earth did not rebound evenly. Rather, large sections slid past other sections, creating vertical fractures that extended to the surface. The fractures initially contained mineral saturated water that, over time, was precipitated to crystalline calcite or dolomite solids.

A 3.5-inch core of the Marcellus Shale was taken by the Department of Energy in 1976 at a location near Morgantown, WV. The drilling location was lucky and intersected a major vertical fracture. The remarkable photograph of the fracture in the core (shown below) was taken by my friend and geoscientist, Bill Overbey.

Vertical fractures, like the one seen in the photograph, extend vertically upward and often are propped open by crystalline calcite or dolomite creating a flow capacity 10 times greater than the shale matrix. The core has a principle vertical fracture ½ inch wide, *plus* a 1/8 inch fault at a right angle to the major fracture, *plus* many small fractures at different angles to the major fracture.

Oil and gas from the shale itself is stored in the fault matrix, making the fractures a primary target for development using horizontal well technology.

Fig. 3 Vertical Fractures in Marcellus Shale Core

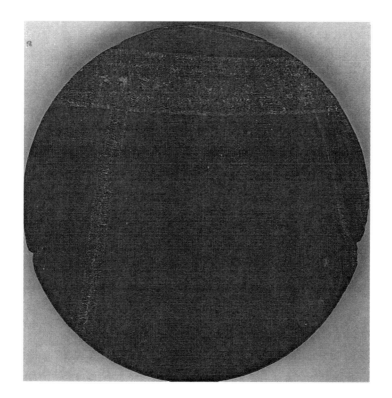

Migrated, Trapped, and Surface Seeps

Oil and gas is forced out of the Marcellus Shale source rock and migrates upward. Little Nell's cousins at the top of the Marcellus, now converted to hydrocarbons, began the journey first. Encountering a cap rock like the Tully Limestone, they become trapped and lay waiting on a drill to liberate them. Once a trap is filled, other hydrocarbon cousins are forced to move elsewhere looking for a path to shallower, more porous formations, where they too become trapped.

Bill Overbey identified twenty-one (21) trap formations holding oil and/or gas that had migrated upward from a source rock at a location near Morgantown, WV.

The Marcellus Shale near Morgantown is 7,400 feet below the Earth's surface and has a thickness of 100 feet (see following figure). The Marcellus is the likely source of the oil and gas pay zones located above it. The Big Injun pay zone is over one mile higher than the Marcellus and once was one of the major gas-producing intervals in WV developed by vertical wells and fracturing.

Fig. 4 Twenty-one Oil and Gas Traps

Expected Pay Zone	Top of Zone, Feet	Zone Thickness, Feet	Indicated Oil or Gas
Big Lime	1180	70	Unkown
Keener, sandy lime	1250	10	Gas
Big Injun limy/dolo./sand	1300	40	Gas
Squaw	1434	14	Gas
Weir	1470	50	Gas & Oil
Berea	1670	10	Gas
Base of Mississippian Upper Devonian			
Gantz	1765	40	Gas
50 Foot	1820	16	Gas
30 Foot	1840	14	Gas
Gordon Stray	1950	12	Gas
2nd Gordon Stray	1980	12	Oil
Gordon	2040	14	Oil
4th Sand	2214	40	Oil - Gas cap
5th Sand	2280	70	Oil
Bayard	2578	22	Gas
Speechley	3010	12	Gas & Oil
Balltown	3550	58	Gas & Oil
Riley	4300	60	Oil - Gas cap
Benson	4640	16	Gas
Geneseo/ Burkett Shale	7150	33	Unkown
Tully Ls.	7183	42	unknown
Mahantango Fm.	7225	175	unknown
Marcellus Shale	7400	100	Gas
Middle Devonian			
Onondaga Limestone	7430	30	unknown
Huntersville Chert	7520	50	Gas
Oriskany Sand	7570	55	Unkown
Helderberg	7940	320	Gas

Other oil and gas traps extend to near the surface, just below the water table. The first well drilled in WV was completed in 1860 at a depth of only 303 feet, in the Cow Run Sand just south of Morgantown. Initial free-flowing production was reported to be about 100 barrels per day, setting off much drilling activity. In 1863, Confederate troops destroyed the oilfield by fire. The burned and abandoned wells allowed fresh water from the water table to enter the formation, watering out previously productive acreage (see *Secondary Oil Recovery Possibilities, Cow Run*

87

Sand by James A. Wasson, Harry R. Johnson, and Dean W. Boley).

Gas is indicated in the figure at two zones below the Marcellus. The likely source of this gas is the Utica Shale that underlies the Marcellus at a maximum depth of about 14,000 feet. The Utica is thicker than the Marcellus (500 feet vs 100 feet) and more geographically extensive. The Marcellus has already become one of the world's largest natural gas fields; the Utica has equal or greater potential for both oil and gas.

But what happened to Little Nell? At the bottom of the pile, she was the last to try to make it to her final resting place. All of the traps were full with her cousins, and she was forced to find a vertical fracture leading to the surface.

She emerged at Oil Creek, PA in 1859 and floated into the atmosphere from a surface oil seep. The seep oil had been collected for years by Seneca Indians and used for ceremonial acts, treatment of stomach ailments, and trading (see *Oil and Gas in Pennsylvania* by Flaherty and Flaherty, starting on page 30). Skimming did not produce enough oil for commercial uses, and investors hired Colonel Drake to bore a hole in the earth near the seep specifically to find and produce oil. Oil was found at 69.5 feet below the surface in August 1859, and signaled the birth of the modern oil industry.

Gas dissolved in shallow drinking water formations is part of the natural process of the upward movement of oil and gas from source rocks. And since source beds are still producing oil and gas, the upward migration will continue into the indefinite future.

Fracture Science

If you remember from chapter 1 of this journey, the Bureau of Mines research and favorable field performance demonstrated that pressure-depleted fields of the Appalachian area would respond to water injection to recover oil. Moreover, the response was predictable, and this permitted a reasonable estimate of project economics. Over time, Appalachian oil producers adopted waterflooding as a standard approach to increase oil recovery from depleted reservoirs, probably by millions of barrels.

However, there were technical issues.

Ohio Clinton sand waterfloods were failing because injected water was moving too quickly along subsurface fractures to a producing well. A lot of oil was being bypassed (see *Clinton Sand Reservoir Characteristics Essential to Successful Waterflooding* by Leo Schrider, 1968). To solve this problem, you need to orient the fractures created by hydraulic fracturing with the natural fractures in the earth.

So how do you do this?

> First, you determine the orientation of the vertical natural fractures using aerial photographs, relief maps, and surface measurements. Bill Overbey and Bob Rough did this and identified the trend for 2,000 vertical fractures from surface outcrops at 16 locations.
>
> Second, you measure the orientation of the fracture caused by hydraulic fracturing of a vertical well (horizontal wells had not yet been invented). For this, you lower a tool equipped with rubber covering and inflate it into the fracture. The fracture imprint retrieved on the

surface gives the subsurface direction of the induced hydraulic fracture as the fracture leaves the wellbore.

Third, you correlate the surface fracture trends with the subsurface hydraulic fracture trends. They were highly correlated, leading to the paper *Surface Studies Predict Orientation of Induced Formation Fractures in Appalachian Area* by William K. Overbey, Jr. and Robert L. Rough, 1968. With this information, oil producers were better able to drill injection and producing well patterns that increased waterflood performance.

Joe Pasini, a petroleum engineer at the Bureau of Mines, reviewed Overbey's ground-breaking work that predicted the orientation of joint systems in the Appalachian Basin. While Overbey was working to solve a technical issue associated with waterflooding, Pasini was interested in increasing primary oil and gas production. He suggested using slant or horizontal wells drilled in the direction needed to intersect the joint systems. Their 1969 Society of Petroleum Engineers paper illustrates in Figure 5 the drilling and fracturing technique to be used (see *Natural and Induced Systems and Their Application to Petroleum Production* by Joe Pasini and Bill Overbey, 1969).

The Society of Petroleum Engineers selected Pasini as a Distinguished Lecturer, and he traveled the U.S. and the world with his message to other petroleum engineers about the value of developing horizontal well technology. The message was received and the first horizontal well was drilled in Texas twenty years later, as detailed in chapter 10.

Fig. 5 Drill Horizontal Wells to Intersect Natural Fractures

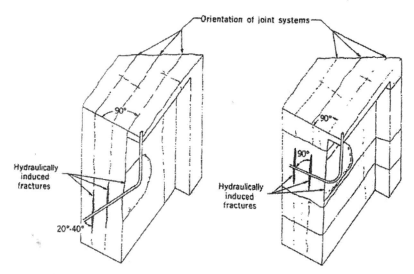

Fig. 7 - Isometric of drilling and fracturing technique to be used in petroleum bearing reservoirs of various thickness and physical properties.

Source: *Natural and Induced Systems and Their Application to Petroleum Production* by J. Pasini III and W.K. Overbey, Society of Petroleum Engineers paper 2565, 1969.

From Fracture Science to Fracture Engineering

Pioneering fracture-science research by Overbey/Rough and Pasini/Overbey:

- *Surface Studies Predict Orientation of Induced Formation Fractures in Appalachian Area* by William K Overbey, Jr. and Robert L. Rough, 1968,

- *Natural and Induced Systems and Their Application to Petroleum Production* by Joe Pasini and Bill Overbey, 1969, and
- *Hydrocarbon Production and Fracture Systems* by Bill Overbey and Joe Pasini, 1970

focused attention on the vast 160,000 square mile Appalachian Basin as a large potential source of natural gas. They estimated that if only 10 percent of the resource is recovered, this would be enough natural gas to supply the eastern U.S. for over 30 years at the then (1977) rate of consumption (see *Natural Gas from Eastern U.S. Shales* by Leo Schrider, Chuck Komar, Joe Pasini, and Bill Overbey, 1977).

However, to achieve that potential, drilling targets within the vast basin needed to be identified. And, once identified, a concentrated effort would be needed to develop the technology necessary to recover the gas. The federal government initiated a major effort, called the Eastern Gas Shales Project, to better characterize the Eastern Shales and to find and test development approaches. By 1978, over thirty projects were underway with a wide range of participants (see *DOE's Unconventional Gas Research Programs*):

Fig. 6 Active DOE Eastern Gas Shales Projects in 1978

Resource Characterization	Extraction Technology
Battelle Ohio	Lawrence Livermore Laboratory
Illinois Geological Survey	Los Alamos Laboratory
Indiana Dept. Natural Resources	Morgantown Energy Tech Center
Juniata College	Petroleum Technology Corp
Morgantown Energy Tech Center	Sandia Laboratory
Mound Laboratory	West Virginia University
Ohio Dept. Natural Resources	
Pennsylvania Geologic Survey	**Technology Testing**
Tennessee Dept. of Conservation	Consolidated Gas
U.S. Geological Survey	*Directionally Drilled Well*
University of Cincinnati	Columbia Gas & Mitchell Energy
University of Kentucky	*Massive Hydraulic Fracturing*
University of North Carolina	
University New York	
West Virginia University	
West Virginia Geological Survey	

Geologic organizations associated with Devonian shale deposits in the Appalachian, Michigan, and Illinois Basins were under contract to characterize and inventory the shales. About 17,000 feet of Devonian shale were collected in nine states for physical and geochemical analysis. Basic geological mapping would be complete by 1980 and compiled, with help from the U.S. Geological Survey, into basin-wide maps to identify regions of high production potential.

Resource characterization data were reviewed by the government's national laboratories and others for ideas about how to stimulate gas production:

- Los Alamos Scientific Laboratory to develop methods for stimulation and characterization of Eastern Shales,
- Sandia Laboratories to determine the orientation and extent of hydraulically-produced underground fractures, and
- Lawrence Livermore Laboratory for predictive methods for massive hydraulic fracturing and related techniques.

By 1978, numerous patents had been issued to government scientists for their pioneering contributions to fracture engineering, including:

- *Horizontal Wells to Remove Methane from Coalbeds*, U.S. Patent 3,934,649 by Joe Pasini and Bill Overbey, 1976,
- *Orienting Induced Fractures in Subterranean Formation*, U.S. Patent 4,005,750 by Zane Shuck, 1977, and
- *Method for Controlled Directional Drilling in Subterranean Earth Formation*, U.S. Patent 4,026,356 by Zane Shuck, 1977.

These and other patents helped assure that the evolving knowledge base was shared widely with other researchers in government and industry. The highlighted patents are referenced even today as others continue to advance horizontal well technology.

From Fracture Engineering to Research Horizontal Wells

The petroleum industry focus was on oil production following the 1978 revolution in Iran (described in chapter 8 of this journey). The industry rapidly advanced horizontal drilling as a new production tool. The initial major target was the Texas Austin Chalk, a naturally-fractured limestone formation. The first successful horizontal well was drilled in 1985, and it was followed by 1,800 other wells by the early 1990s (see *Horizontal Drilling in Deep Austin Chalk*, 1997). The horizontal wells were drilled perpendicular to the natural fractures, just as Pasini and Overbey had suggested twenty years earlier.

Geologic data from the Eastern Gas Shales project were used to site the 1986 Department of Energy's Devonian shale horizontal well in Wayne County, West Virginia. This research well was drilled to a vertical depth of 3,000 feet, cased, and cemented to the surface. Then, air was used to drill a horizontal hole of 2,000 feet, setting a world record for an air-drilled horizontal well. A video camera survey recorded the location of over 200 natural fractures, many of which were observed bubbling natural gas liquids. The 2,000 foot long horizontal section was cased, but not cemented to avoid water contact with the shale. Eleven (11) fracture tests were performed and the results measured in terms of production improvement. Only gas fractures (no water) were used at the DOE research well:

- Nitrogen gas,
- Liquid CO_2, and
- Nitrogen foam with proppant

From Research Horizontal to Commercial Horizontal Wells

To help move the technology from research to commercial operations, the DOE then cost-shared two horizontal wells with industry partners, Cabot Oil and Gas and Consolidated Natural Gas. Martin Adams and Bennie DiBona had both returned to consulting from their days of government service. They had won the contract to support drilling and completing two commercial horizontal wells, and had negotiated terms with two industry partners.

Each well was drilled, and the horizontal well completed using the same approach as the 1986 DOE research well. To avoid water touching the formation to be fractured, the horizontal length of the well was drilled using air instead of mud to lift the cuttings to the surface. The horizontal well was cased but not cemented. Fracing used nitrogen foam with proppant; no water was used to create these fractures. Each well was placed on production by 1991.

The DOE horizontal wells and related work are well documented in the professional literature with complete citations provided in the references to chapter 9. Information in the reports provides a valuable guide to re-examining completion approaches that do not require water as the fracing fluid:

- *Devonian Shale Horizontal Well: Rationale for Wellsite Selection and Well Design*
- *Drilling a 2,000-ft Horizontal Well in the Devonian Shale*
- *Analysis of Natural Fractures Observed by Borehole Video Camera in a Horizontal Well*
- *Inducing Multiple Hydraulic Fractures from a Horizontal Wellbore*
- *Site Selection, Drilling, and Completion of Two Horizontal Wells in WV*
- *Liquid-Free CO2/Sand Stimulations: An Overlooked Technology*

Interest in Devonian shale development lay dormant for the next decade as the industry continued to apply horizontal technology to other locations, such as the Bakken Shale of North Dakota and the Barnett Shale of Texas.

George Mitchell is given credit for unlocking the Barnett Shale using massive hydraulic fracturing, summarized as follows from *The Boom in Shale gas? Credit the feds*, by Michael Shellenberger and Ted Nordhaus, Wall Street J., Dec 17, 2011:

> "Slick-water fracking, the technology that Mitchell used to crack the shale gas code, was adapted from massive hydraulic fracturing, a technology first demonstrated by the Energy Department in 1977 (Note: As shown in Fig. 6, Mitchell Energy was a subcontractor to Columbia Gas testing massive hydraulic fracturing). Over the next two decades, Mitchell and others, with government support, tinkered with the technology, exploring ways to use fewer chemicals and more water, which substantially reduced the cost of extraction.
>
> Mitchell learned of shale's potential from the Eastern Gas Shales Project, a partnership begun in 1976 between the Energy Department's Morgantown Energy Research Center and dozens of companies and universities that sought to demonstrate natural gas recovery in shale formations and to map and test core samples from unconventional natural gas deposits. Starting in 1981, Mitchell's geologists drew heavily on that research to guide their explorations."

Range Resource geologists noted that the Marcellus Shale of the Appalachian Basin has properties similar to those of the Barnett Shale of West Texas, which was a prolific source of natural gas. In 2004, Range drilled its first horizontal well in the Marcellus. The well was completed in Pennsylvania using the massive hydraulic fracing techniques developed for the Barnett. The well was placed on production in 2005, setting off a drilling boom. By

2010, 1,446 Marcellus horizontal wells had been drilled in Pennsylvania (see *Marcellus Natural Gas Trend* from Wikipedia).

The large increase in production from the Marcellus Shale has pushed total United States gas production to a new all-time high, as shown in the next figure:

Fig. 7 United States Gas Production Reaches All-Time High

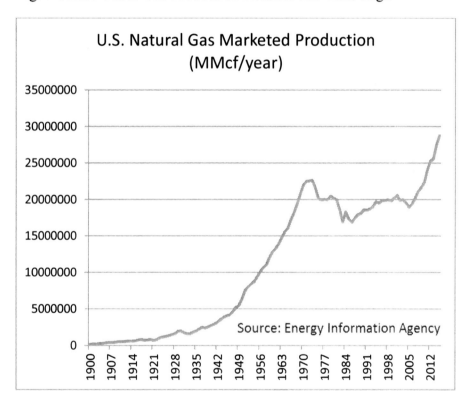

The Marcellus is now considered to hold the largest volume of recoverable gas in the U.S. Even so, it may soon be replaced with gas from the underlying Utica Shale formation. Gas production from the Utica is growing rapidly, perhaps helped with a comprehensive update of the geology lead by my friend Doug Patchen of West Virginia University (see *A Geologic Play Book for Trenton-Black River Appalachian Basin Exploration* by Doug Patchen and others, including the Geological Surveys of Ohio, Pennsylvania, and West Virginia). Together, both the Marcellus and Utica shale formations hold an important key in our journey toward energy independence.

Once again, another 1977 MOPPS finding (see Chapter 6) has been validated:

> "Huge quantities of natural gas could be unleashed by using conventional and some advanced technologies (horizontal drilling and simulative fracture) if wellhead prices increased to more realistic levels. Natural gas sufficient to meet the nation's needs for well over a century would be forthcoming..."

But at what cost?

Water Management in an Environmental Age

Water used in fracing and water returned to the surface from fracturing operations depend on location. The Environmental Protection Agency (EPA) gives two examples (see *Assessment of the Potential Impacts of Hydraulic Fracturing for Oil and Gas on Drinking Water Sources*, Figure ES-3).

Marcellus shale wells, according to the EPA example, typically use 4.6 million gallons of water for fracing a single horizontal well. The water dissolves minerals in the vertical fractures and returns to the surface with a salt content of up to 200,000 parts per million, about 5 times saltier than sea water. Over the life of the well, 2.3 million barrels of salty water will be produced for disposal. The water will be separated from the gas and stored. When the water storage units are full, the salty water will be picked up by a truck and hauled for disposal.

Barnett Shale wells in Texas, according to the EPA example, use a volume of water similar to the Marcellus in fracture operations: 4.5 million gallons. None of the water is retained in the reservoir, since each well returns 4.5 million gallons to the surface. Only 5 percent of the water can be recycled; the other 95 percent (4.3 million gallons) would be sent for disposal.

From these EPA examples, horizontal wells will return from 2.3 million gallons to 4.3 million gallons of salty water for disposal.

Disposal where?

Salt water disposal wells. The nation currently has some 144,000 wells that receive 47 million barrels of salty water each day; on average, 326 barrels per day per well.

Increasing amounts of water produced with oil and gas production have forced the industry to increase both the rate of injection and the injection pressure. Seismic events (mini earthquakes) have long been linked with oil and/or gas water disposal operations ranging from Arkansas, to Texas, to Colorado, and to Ohio. The epicenter is now northern Oklahoma, where injection-linked earthquakes have forced the shut-ins of hundreds of water disposal sites. In 2016, Oklahoma experienced three (3) earthquakes with a magnitude of 5.0 or more; the rest of the U.S. had only two (2) such quakes.

Water disposal problems will only get worse.

Grady County, near Oklahoma City, is one of the most active development areas in the state. In 2014, 80 million barrels of salt water were produced, or 222,000 barrels per day (see *More Oil, More Water*, Journal of Petroleum Technology, Dec. 2016). Only 3 percent of 80 million is needed to support the entire oil and gas operations in Grady County.

What about the 97 percent not needed?

The industry is building the infrastructure to move the salt water by truck or by pipeline to other disposal locations. This could work if water is moved far enough, and if existing wells are converted to disposal or new wells are drilled. Excess water from Grady County alone will require 600 to 700 disposal wells.

Another approach is to boil the water, leaving behind the salt for disposal. For Grady County, this will require 16,500 railcars each year. About 1 railcar every 30 minutes would need to be filled, and the salt moved somewhere. Where can you move that much salt where it can be protected from rain that would remobilize it into the environment?

Water management is becoming big business. Boiling the water in Grady County alone will cost about $400 million each year, or billions per year if water management in other development areas is included.

Water management may limit the amount of oil and/or gas that can safely be produced in the United States. The petroleum industry may then turn to the only long-term alternative: eliminate water in fracing operations. A concerted effort like the Eastern Gas Shales Project will be needed to fully develop gas (liquefied natural gas, nitrogen, or carbon dioxide) as the fracture fluid alternative to water.

Chapter 10

UNCONVENTIONAL OIL

OIL IMPORTS WERE ONCE AGAIN threatening the nation's economy in 2004. This is what the nation was facing:

Fig. 8 United States Oil Imports Reach Record High in 2004

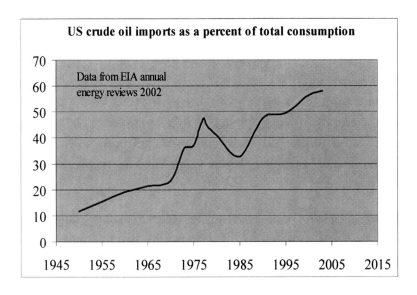

Source: *Strategic Significance of America's Oil Shale Resources,* 2004

The U.S. responded to energy threats in the 1970s with regulation, funding, tax incentives, and executive orders (detailed in chapter 8).

By throwing the kitchen sink at this problem, we were able to increase oil production by 10 percent and cut oil imports from 48 percent to 32 percent of total consumption.

In response, the Organization of Petroleum Exporting Countries (OPEC) began to flood the world market with oil. Oil prices began a dramatic decline, driving U.S. oil producers out of business and banks to failure. Lower domestic oil production forced oil refiners to buy foreign crude, driving up oil imports.

Oil imports were forecast by the government to exceed 70 percent of demand by 2025, with the vast majority coming from OPEC nations. America's vulnerability to price shocks, disruptions, and shortages would also increase. The U.S. was again considering a massive synfuels program that was tried, without success, in the 1980s.

Initial Focus on Oil shale

Liquids from the nation's vast oil shale deposits once again emerged as a key target for development. Nearly two decades had passed since meaningful federal oil shale initiatives had been taken. Since that time, oil shale technology had advanced and the regulatory landscape had matured.

Because of my prior government roles managing large oil shale studies, I was recruited to collaborate with Peter Crawford on an oil shale update for the federal government as an employee of INTEK, Inc., an engineering consulting firm started by my friend from Oklahoma, Khosrow Biglarbigi, and his insightful engineering/mathematician partner Hitesh Mohan.

We produced by March 2004 *Strategic Significance of America's Oil Shale Resources,* published as two volumes: *V. 1 Assessment of Strategic Issues* and *V. 2 Oil Shale Resources Technology and*

Economics by Harry R. Johnson, Peter M. Crawford, and James W. Bunger.

Preparation of this report proved to be the turning point in the careers of three friends. Khosrow Biglarbigi, with his photographic memory, became an expert in oil shale technology. As a distinguished lecturer for the Society of Petroleum Engineers, he traveled the country and the world with his message about the *Potential for Oil Shale Development in the United States*, 2007. He later designed with his mining engineer partner, Hitesh Mohan, and Peter Crawford a commercial overseas oil shale facility, including siting, mining, retorting, and upgrading the raw produced shale. Peter Crawford became the primary INTEK support to the Department of Energy in its renewed efforts to get liquids from oil shale. His continuing contributions are noted in the publication credits that follow. The oil shale baton had been successfully passed to three talented associates.

Our two-volume analysis attracted widespread interest and favorable comments. Continuing with these planning efforts, Anton R. Dammer, Director of Naval Petroleum and Oil Shale Reserves, chartered a study to guide federal actions designed to accelerate private industry development of a domestic oil shale industry. *America's Oil Shale A Roadmap for Federal Decision Making* by James W. Bunger, Peter M. Crawford, and Harry R. Johnson was published in Dec. 2004.

Congress, in the *Energy Policy Act of 2005*, then directed the Secretary of Energy to establish a Task Force to:

> ...develop a program to coordinate and accelerate the commercial development of strategic unconventional fuels, including, but not limited to, oil shale...

New Massive Synfuels Program Being Developed

The Congressional directive set in motion efforts to develop a new massive synfuels program. The required task force was formed from federal, state, and local organizations. It consisted of three Cabinet organizations (Secretaries of the Departments of Energy, Defense, and Interior), five state organizations (governors of Colorado, Kentucky, Mississippi, Utah, and Wyoming), and representatives from local communities that would be impacted by development. The *Task Force on Strategic Unconventional Fuels*, supported by Peter Crawford, was published in September 2007.

When the task force study was released, 32 private companies were actively researching new and/or improved approaches for oil shale and heavy oil development. Laboratory and field activities of these companies were compiled by Peter Crawford and first published in June 2007 as *Profiles of Companies Engaged in Domestic Oil Shale and Tar Sands Resource and Technology Development*.

Private and political momentum was now in place to trigger federal actions that would lead to a massive synfuels program. Would this be the response to James W. Bunger, Peter M. Crawford, and Harry R. Johnson's Oil & Gas Journal article *Is Oil Shale America's Answer to Peak-Oil Challenge?*

Not this time.

Horizontal Wells to the Rescue

Texas led the way with the first successful horizontal well drilled in 1985 in the Austin Chalk. It was followed by 1,800 horizontal wells by the early 1990s. Texas oil production began to stabilize, then rise. The new technology spread to other unconventional deposits. The first Marcellus shale horizontal well was placed on production in 2005, setting off a drilling boom. By 2010, 1,446 new wells had been drilled in Pennsylvania. In 2006, a single horizontal well drilled into an oil-rich layer of the Bakken Shale of North Dakota ignited intense development. In just four years, oil production reached 458,000 barrels per day, outstripping the capacity of the pipeline to ship oil out of the region.

The Texas horizontal well and massive hydraulic fracturing technology stabilized, and then began to increase overall U.S. crude oil production (see Figure). By 2016, oil production in the U.S. was near its 1970 peak of about 10 million barrels per day, as shown below:

Fig. 9. Unconventional Oil Production Revives U.S. Petroleum Industry

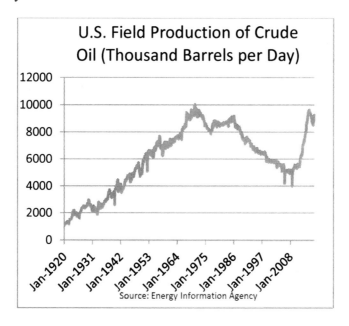

What a remarkable effort by the petroleum industry! They reversed the 40-year oil production decline from 1970 to 2010, and it only took about a decade to accomplish!

Political leaders followed these developments. Plans for a massive government synfuels program were abandoned.

OPEC followed these developments. Increased U.S. production again threatened their share of world markets. Saudi Arabia began a two-year program to protect its market share by flooding the market with cheap oil.

Two years of low oil prices had its impact (see *In Saudi-Shale Fight, Both Claim Victory*, Wall Street J, Dec. 16, 2016). Exploration and production company bankruptcies grew to 100. Spending on new projects fell dramatically, from $21 billion in 2014 to $4 billion in 2016. Drilling rigs became stacked. More than 100,000 workers lost their jobs. Producers began to shut in oil production. New horizontal wells were not completed with massive hydraulic fractures. But overall oil production fell only about 10 percent.

Flooding the market with cheap oil worked in the 1980s. This time was different. This time, U.S. producers had shifted from drilling vertical wells into depleting conventional oil deposits to drilling horizontal wells into unlimited unconventional deposits

Because of this shift from conventional to unconventional oil deposits, the U.S. was able to maintain a near historical level of oil production. Producers also had significant shut-in production potential that could be tapped at the right price.

Saudi Arabia knew its two-year effort to flood the market with cheap oil was not working. They began to work with other world producers to reduce oil production and increase oil prices to a more sustainable level. The OPEC and Russian oil production cuts began in early 2017.

U.S. oil production began a steady increase and, between November 2016 and April 2017, production rose by 150,000 barrels per day each month. By April 2017, oil production was over 9 million barrels per day and still climbing (see Figure 9).

The U.S. is beginning to emerge as the world's new swing oil producer.

Chapter 11

UNITED STATES RESPONDS TO LIBYAN CIVIL WAR

A CIVIL WAR in Libya began in February 2011 when security forces fired on a crowd seeking to oust the government of Colonel Muammar Gaddafi. Protests escalated into a rebellion that spread across the country.

The Libyan oil industry infrastructure was heavily damaged. Inland pipelines and pump stations from major oil fields were destroyed. Shipping terminals along the Mediterranean Sea were damaged. In just three (3) months, 132 million barrels of crude from Libya were lost to world markets.

The major impact was at European refineries, where an oil storage surplus of 31 million barrels fell to a 44 million barrel deficit. Loaded oil tankers were diverted from the United States to Europe refiners, lowering oil stocks in the United States. Saudi Arabia's oil production surged by 1 million barrels per day to a 30-year high of 9.7 million barrels per day.

Despite the worldwide efforts to manage the loss of Libyan crude, it became clear that additional efforts were needed. The International Energy Agency (IEA) decided to act and announced plans on June 23, 2011, to release 60 million barrels of oil and products from the stockpiles of member countries:

- 30 million barrels were released from locations in the Pacific Region (Japan and Korea) and the Europe Region (France, U.K. Germany, Italy Spain, and others); and
- 30 million barrels were released from the U.S. Strategic Petroleum Reserve (SPR).

The SPR was at last ready, a far cry from the two earlier embargos:

> The Arab oil embargo of 1973 (see Chapter 5) caused oil prices to jump 250 percent almost overnight. Gasoline and diesel shortages rapidly appeared. Motorists waited in long lines for expensive, limited amounts of gasoline. Truckers blocked traffic on major highways.

The economic shockwaves pushed the nation into plans that would mitigate future energy emergencies. Specifically, the *Energy Policy and Conservation Act of 1975* declared it to be U.S. policy to establish a reserve of petroleum, setting the creation of the SPR in motion.

> Construction of the SPR was underway in 1978 when the civil war in Iran erupted, cutting off all exports from that country (see Chapter 8). Shortages of gasoline and diesel once again appeared across the nation. Oil, gasoline, and diesel prices doubled overnight. Long lines for gasoline returned. Truckers began to strike and again block traffic at vital locations. There was indeed *Panic at the Pump* (see book by Meg Jacobs).

Congress directed that the SPR fill rate be accelerated with passage of the *Energy Security Act of 1980*, signed into law by

President Jimmy Carter. Two years later, President Ronald Regan signed the *Energy Emergency Preparedness Act of 1982* that further accelerated the SPR fill rate and increased its size to a minimum of at least 500 million barrels in storage.

The SPR continued to be filled and stored in enormous caverns mined from naturally occurring salt formations along the Gulf of Mexico in Texas and Louisiana. Each cavern can hold a nominal 10 million barrels of oil — some more, some less. Drawdown capability is 4.4 million barrels per day, limited by the capacity of the pipelines and marine terminals that move oil from cavern storage to refiners.

Strategic Petroleum Reserve Crude Compatibility Evaluated

Many refineries along the Gulf Coast had installed sophisticated equipment to refine increasing amounts of heavy oil coming from Mexico and Venezuela. In contrast, the oil stored in the SPR is high quality light oil. So the ability to refine SPR oil was the question asked by Dave Johnson, Deputy Assistant Secretary for Petroleum Reserves, as part of his efforts to improve SPR capabilities. Specifically, he asked if the nation's refineries could use the oil stored in the SPR without adversely impacting the refinery product slate of products.

To answer that question, the INTEK team led by me and my old friend from Morgantown, WV, Joe Pasini, developed an engineering refinery model to evaluate the limits of individual refineries to substitute SPR crude in place of their usual foreign crude supplies (see *Strategic Petroleum Reserve Crude Compatibility Study*, December 2005). Of the nation's 149 refineries operational in 2004, 11 refineries had a low capability to substitute heavy oil from Mexico and Venezuela for SPR oil. If all of these imports were disrupted, these 11 refineries could still use SPR crude, but would need to reduce the amount of oil processed.

Our analysis was about to be tested.

Strategic Petroleum Reserve 2011 Libyan Drawdown

By the 2011 Libyan oil disruption, the SPR had over 700 million barrels in storage and was ready to respond. Detailed planning for a large drawdown from the Reserve was under the able direction of David Johnson. Plans were in place and practiced that allowed the 30 million barrel drawdown to proceed without incident.

The June 24, 2011, directive from the President set off a complicated series of actions that required the DOE to offer for sale oil stored at SPR locations. The offer was oversubscribed by an industry hungry for crude. Contracts were awarded by the DOE; oil deliveries began just 21 days after the President's directive.

Pipelines connected to SPR sites moved 21 percent of the oil (see *Strategic Petroleum Reserve Annual Report for Calendar Year 2011)*. The balance was moved largely by tanker. 77 percent of the oil was delivered to the nation's largest concentration of refineries along the Gulf of Mexico. However, refineries along both coasts of the U.S. received oil. Once loaded, tankers could reach U.S. Virgin Islands refiners in a few days and East Coast refiners (Pennsylvania, New Jersey, and Delaware) in about a week. Other tankers carried oil through the Panama Canal for delivery to California and Washington in two weeks, and Hawaii in three weeks.

Oil released from the SPR entered the market in July and August. 30.64 million barrels were delivered by August 31, 2011, as originally scheduled.

In September 2011, the IEA concluded:

> "...that the interrupted Libyan supplies have been successful addressed by a combination of the IEA collective action and increased production from producer countries..."

The SPR worked as designed, and all of the oil released was refined without difficulty, validating our crude compatibility analysis.

The Strategic Petroleum Reserve now stands ready as an effective short-term shield against future supply interruptions.

Chapter 12

COMPLETING THE JOURNEY TOWARD ENERGY INDEPENDENCE

Well, it's been quite a journey, but the end is in sight. This final chapter traces the flow of oil imports that have crippled the nation in the past. In response, the U.S. constructed and filled the Strategic Petroleum Reserve that has proven to be an effective shield against oil supply interruptions. Finally, the vast North America shale deposits, coupled with horizontal wells and massive hydraulic fracturing, have made energy independence possible. Let's start with the nation's relationship with imported oil over the past six decades.

Flow of Oil Imports Since 1960

Early in this period, the United States lost its place as the world's largest oil producer to the Organization of Petroleum Exporting Countries (OPEC). Using its new powers, OPEC punished us for our 1973 support of Israel with a total oil embargo, leading to long lines to get gasoline. Five years later, in 1978, the revolution in Iran cut off all oil exports, leading again to long lines at our gasoline pumps and a demand for government action.

My job was to help develop and manage government programs that increased domestic oil production by 10 percent between 1976 and 1985. Increased production lowered our dependence on imported oil, but OPEC saw it as a threat to its dominance of the world oil market. OPEC reacted by flooding the world with cheap oil. Domestic oil producers could not compete. Many went bankrupt, while large banks failed and domestic oil production fell for the next 30 years. OPEC had won.

The Eastern Gas Shales Project was chartered by the government to find ways to extract oil and gas from the Devonian and other eastern shales. By 1978, over thirty projects were underway by a variety of organizations: federal and state geologic surveys, government national laboratories, and private oil and gas producers. The program focused attention on the vast oil and gas potential of the shales and created industry interest in developing horizontal well technology.

The industry developed and drilled the first commercial horizontal well in Texas in 1985. It was drilled perpendicular to the natural fractures, just as government researchers had earlier recommended. Horizontal well technology and massive hydraulic fracing spread rapidly. Today, the Marcellus shale provides over 20 percent of the total gas produced in the United States, and it is considered to hold the nation's largest volume of recoverable gas.

Oil produced using horizontal well technology reversed a 40-year decline in oil production. By 2014, U.S. oil production was near its all-time 1970 production peak. OPEC was again losing market share and reacted by again flooding the world with cheap oil. U.S. producers bent, but did not break. New well investments fell, drilling rigs became stacked, workers lost jobs, and producing wells were shut in. Even with these measures, domestic oil production remained within 10 percent of the all-time peak production set in 1970.

The OPEC effort to flood the world with cheap oil did not work this time. They gave up after two years and, in 2016, began to reduce their oil production to increase oil prices to a more sustainable level. OPEC had lost.

The U.S. is now emerging as the most important oil and gas producer in the world.

So what's next?

Effective Response to Oil Import Disruptions Developed

The Arab oil embargo of 1973 sent shock waves throughout the United States. Oil prices jumped 250 percent in a few months, gasoline and diesel shortages rapidly appeared, motorists waited in long lines for gasoline, truckers blocked traffic on major highways, half a million people lost their jobs, and consumer prices skyrocketed. Congress decided to build a Strategic Petroleum Reserve to guard against future shortfalls.

The Iran civil war of 1978 found the SPR under construction and not yet ready for prime time. Gasoline lines returned, truckers blocked traffic, and things got ugly with local riots. The government was forced to act and started programs that could have cost nearly $1 trillion. Some programs were effective in increasing domestic oil production. Protecting their interests, Arab cheap oil flooded the world market and killed the recovery of our petroleum industry.

Oil imports reached nearly 60 percent of consumption by 2004, and imports were forecast to continue to increase. The nation prepared for another massive intervention to increase oil production from oil shale, tar, and coal; a synfuels program that could have cost $1 trillion or more.

Horizontal drilling bailed us out. Oil produced from unconventional formations reversed a 40-year decline from conventional formations. Plans for a massive government synfuels program were abandoned.

By 2011, the SPR had over 700 million barrels in storage and could move large quantities of oil by pipeline and tankers. A worldwide shortfall of oil from a civil war in Libya was easily offset by coordinated oil and product releases from International Energy Agency (IEA) nations, including the United States. The Strategic Petroleum Reserve (SPR) began to deliver oil only 21 days after the President's drawdown directive.

The U.S. is far less vulnerable to interruptions in oil supplies than ever before. The SPR is now proven to be an effective short-term shield against supply interruptions. The International Energy Agency has proven to be a reliable source of stored oil and products.

Domestically, shut-in production from horizontal wells could be opened as rising oil prices caused by a supply shortfall make production profitable. More wells could be added, depending on the expected future oil price.

Additionally, China has constructed its own SPR and has filled it with cheap oil over the years. The planned amount of oil in storage is 475 million barrels. India is also building an SPR and has 36 million barrels in storage. During a supply interruption, China and India drawdowns would ease the supply demands from IEA nations.

The world has developed an effective response to oil supply interruptions. Now the United States is beginning to turn to energy independence.

Massive Unconventional Shale Deposits

Organic deposits similar to those found in the Appalachian Basin are widely distributed within the United States shale basins. Current developments and prospective future developments are displayed in the following figure (reproduced from *Lower 48 states shale plays* by Energy Information Agency). The diversity and size of the deposits provide the resource base needed to achieve energy independent.

Figure 10 Massive Unconventional Shale Deposits Widely Dispersed Across U.S

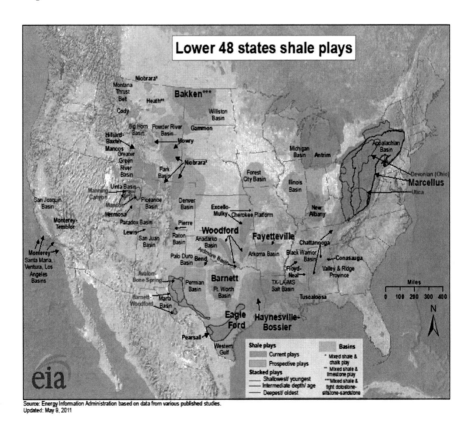

One major feature in the figure is the size of the Appalachian basin that stretches north-south from New York to Kentucky and east-west from Maryland to Ohio. Over 160,000 square miles in area, this deposit is the home of both the Marcellus and Utica shales, deposited in an ancient Devonian sea discussed in chapter 9 (remember Little Nell?).

Forty years ago, in 1977, I testified at a House of Representatives hearing (see chapter 6) that:

> "The Devonian shale might be about 50 trillion Cubic Feet (Tcf), again a large resource that underlies most of the East Coast...this is an area that we are exploring and delineating the resource right now..."

That was a low estimate. Recoverable gas in the Marcellus plus Utica shale is now estimated by the EIA to be 402 Tcf, and the estimate continues to grow based on industry field drilling and results. Even so, the current recoverable gas from just these two shale deposits is large enough to supply the needs of the entire nation for 15 years at our 2015 rate of consumption of 27.1 Tcf.

The Bakken shale was deposited in the Devonian Age, about 400 million years ago, or about the same time as the Marcellus shale was deposited on our east coast. This formation underlies parts of Montana, North Dakota, Saskatchewan, and Manitoba. The pressure and heat were enough to cook the decaying animals and plants into oil, but were not sufficient to convert them to gas. Therefore, the Bakken mostly produces oil and natural gas liquids, while the Marcellus and Utica mostly produce natural gas.

The amount of technically recoverable oil and gas is staggering. In the U.S. alone, EIA estimates recoverable liquids total 135.8

billion barrels, or 40 years of supply at our 3.4 billion rate of production in 2015. Similarly, recoverable gas is estimated at 1,075 Tcf, or 40 years of supply at our 27.1 Tcf rate of production in 2015 (see Table 9.3 in *Assumptions to the Annual Energy Outlook 2016* by EIA). And the recoverable volumes continue to grow as drilling better defines potentials.

Shale deposits do not end at the political borders shown in the figure.

To the south, on the U.S. side, EIA estimates the Eagle Ford shale of Texas to have 21.8 billion barrels recoverable. On the Mexican side, the Eagle Ford has an additional 8.1 billion barrels.

To the north, on the U.S. side, EIA estimates the Bakken shale of North Dakota to have 25.6 billion barrels technically recoverable. On the Canadian side, the Bakken has an additional 8.8 billion barrels.

Canada has other vast deposits, the largest being the Horn River Shale that lies east of the Rocky Mountains in Alberta and British Columbia. Up to 1,000 feet in thickness, this Devonian Age deposit has a natural gas potential that rivals the gas potential of the Marcellus plus Utica Devonian Age deposits of the Appalachian Basin. In addition, the natural gas liquid potential is double the size of the Canadian Bakken deposit. Truly a giant among giants.

The shale resource base of North America coupled with horizontal wells and massive hydraulic fracturing has made energy independence possible. Sometime in 2017, the U.S. will likely sell more natural gas to the world than we buy. Additionally, the EIA expects total imports to the U.S. to be balanced with exports from the U.S. by 2026.

When oil and gas exports balance oil and gas imports, the nation will have completed its journey toward energy independence, probably within the next ten years.

We have one remaining problem to address – the disposal of produced water.

More Oil More Water

Each horizontal well massively hydraulically fractured will return 2.3 million to 4.5 million gallons of water to the surface for disposal (see chapter 9 of this journey). Moreover, the returning salt water will have a salt content of up to 200,000 parts per million, about 5 times saltier than sea water.

One solution is to boil the water and move the residual salt for disposal. In one small county in Oklahoma, the remaining salt will fill 16,500 railcars each year. Every 30 minutes, one railcar would be loaded and the salt moved to a disposal location. How do you protect this volume of salt from being remobilized into the environment by rain?

The current solution is salt water disposal wells. The U.S. has some 144,000 disposal wells that receive 47 million barrels per day of salty water. Increasing amounts of salty water have forced the industry to increase both the rate of water injected and the injection pressure. This has caused mini earthquakes that are associated with oil and gas disposal operations. Oklahoma used to have one or two seismic events a year; they now have at least one each day. The quake severity is also increasing. In 2016, Oklahoma had three large quakes with a magnitude of 5.0 or more, while the rest of the U.S. had only two such quakes. The state has already closed hundreds of water disposal sites, and more may follow.

Management of excess salty water may limit the amount of oil and/or gas that can safely be produced in the United States and impact our journey to energy independence.

Substituting gas for water in fracturing operations

may be the only long-term solution.

The Department of Energy fractured its first Devonian shale horizontal well in 1986 without using water. This was followed by two horizontal wells cost-shared with industry partners. Only gas fractures (nitrogen, liquid CO_2, and nitrogen foam with proppant) were used in completing each of these wells. No water was used to create fractures in the shale.

Each of the three horizontal wells was drilled at a depth of about 3,500 feet. Current horizontal wells in the Marcellus are at 7,400 feet. The pressure and the formation heat at this depth will pose significant technology challenges that must be addressed. It is not likely than any one commercial organization will drill and fracture deep horizontal wells with gas without a full understanding of the technical, cost, and environmental issues.

In the thoughtful editorial published in the Washington Post (*The boom in shale gas? Credit the feds*, by Michael Shellenberger and Ted Nordhaus, Dec 17, 2011), the authors conclude that:

> "The breakthroughs that revolutionized the natural gas industry – massive hydraulic fracturing, new mapping tools and horizontal drilling – were made possible by the government agencies that critics insist are incapable of investing wisely in new technology.

…there's no denying the extraordinary economic return on taxpayer investments. Shale gas is likely to allow the United States to go from net gas importer to a net gas exporter over the next decade.

…private sector alone cannot sustain the kind of long-term investments necessary for big technological breakthroughs in the midst of volatile energy markets and short-term pressure to produce profits.

No doubt, government energy innovation investments could be made more efficiently and effectively. But it would be a mistake to imagine that we'd better off without them."

A highly focused effort like the Eastern Gas Shales Project described in chapter 9 of this journey is needed to fully develop gas as the alternative to water. The new project, like before, would require geologic expertise, laboratory research, evaluation of alternative fracturing gases, experimental wells to test high potential approaches, and cost-shared commercial wells to establish economics. The project will take time but could pave big dividends as the nation continues its Journey Toward Energy Independence.

Acknowledgements

BILL OVERBEY AND JOE PASINI were lifelong friends and associates who provided the technical foundation for the nation's Journey Toward Energy Independence. Bill learned how to predict the orientation of natural fractures in the earth's crust. Joe used this information to suggest horizontal wells be drilled through the fractures to increase oil and gas recovery. They obtained the first horizontal well patent. They promoted horizontal well technology throughout the United States at meetings of the Society of Petroleum Engineers and other scientific organizations. I acknowledge with respect the contributions of Overbey and Pasini leading to industry interest in developing and applying horizontal technology to increase oil and gas from the nation's vast unconventional resources.

With the passing of both Overbey and Pasini, I recruited expert geologist Doug Patchen from West Virginia University to review what I said about the geology of the unconventional resources of the Appalachian basin. He not only improved the geologic description, he twice provided detailed editorial reviews of the text. For both, I am grateful.

Peter Crawford, an excellent organizer and author of technical material from INTEK, Inc., helped me structure the book and provided useful suggestions about content. Bob Folstein, my deputy from our days running the Bartlesville (OK) Energy

Technology Center, is now the editor of the Jacque Offenbach Society quarterly newsletter. He treated me as one of his newsletter's authors, providing several insightful reviews of my book as it was being prepared. Thanks to both for your helpful comments and insights.

Khosrow Biglarbigi, now president of INTEK, Inc., said he enjoyed reviewing the text but found a hole in the government incentive packages designed to increase oil and gas production. He and Peter Crawford provided extensive information about state tax incentives that were adopted by oil producing states, which are summarized in the text of the book. Thanks for the contribution.

The Market Oriented Program Study (MOPPS) was a major event in my career that led to increased funding for oil and gas research and, ultimately, the technical foundation for the horizontal technology now being applied to our vast unconventional deposits. Martin Adams was the chief architect of the study and provided several insightful reviews of the text as it was being developed. Thank you, Martin, I enjoyed working with you once again.

The text cites over 70 individuals who have contributed to this *Journey*. These include co-authors on technical papers, my bosses who provided mentoring and guidance, personal meetings with heads of government agencies (Secretary of Interior, Administrator of the Energy Research and Development Administration, Administrator of the Federal Energy Agency, Secretary of Energy, and President of the United States), and testimony before the House of Representatives and the United States Senate.

And finally, a big thanks to my wife of now nearly six decades, Louise. She reviewed each chapter as it was drafted and offered useful comments from a non-technical perspective. She thought I was a little harsh on Little Nell, but heck; 400 million years ago, everything died and floated to the bottom of the ocean.

Our two hillbilly kids are now full grown and happy. Ray moved with us from Maryland to Oklahoma and decided to stay in Indian country. Leslie is in the process of moving from Pennsylvania to northern Virginia, I think to make sure we continue to take our meds as we grow older.

To all, I appreciated the insights offered and the chance to once again relive old memories. Thank you.

References

Chapter 1. Waterflooding Appalachian Oilfields

Waterflood Possibilities of the Clinton Sand, Logan Oilfield, Hocking County, Ohio by Harry R. Johnson and Dean W. Boley, Producers Monthly, V. 27, No. 12, December 1963.

Predicted Oil recovery from a Pilot Waterflood in the Kane Oilfield, Elk County, PA by Dean W. Boley, Harry R. Johnson, and John R. Duda, Producers Monthly, v. 28, No. 8, August 1964.

Secondary Oil Recovery Possibilities Cow Run Sand, Burning Springs Pool, Wirt County, W. Va. By James A. Wasson, Harry R. Johnson, and Dean W. Boley, Bureau of Mines Report of Investigations 6460, 1964.

Oil Reservoir Analysis and Predicted Recovery by Waterflooding, Clinton Sand, Logan Oilfield, Hocking County, Ohio by D. W. Boley, H. R. Johnson, and W. K. Oberbey, Jr., Bureau of Mines Report of Investigations 6683, 1965.

Oil Recovery by Low-Pressure Gas Drive in the Keener Sand, Bonds Creek Oilfield, Lafayette District, Pleasants County, W. VA by Leo A. Schrider and James A. Wasson, Bureau of Mines Report of Investigations 6798, 1966

Predicted Oil Recovery by Waterflood and Gas Drive, Bradford Third and Sartwell Sands, Sartwell Oilfield, McKean County, PA by John R. Duda, William K. Oberbey, Jr., and Harry R. Johnson, Bureau of Mines Report of Investigations 6943, 1967.

Theoretical and Field Waterflood Performance, Kane Sand, Kane Oilfield, Elk County, PA by Leo A. Schrider, John R. Duda, and Harry R. Johnson, Bureau of Mines Report of Investigations 6917, 1967.

Carbon Deposition for Thermal Recovery of Petroleum; A Statistical Approach to Research by Harry R. Johnson and Edward L. Burwell, Bureau of Mines Report of Investigations 6756, 1966 and reprinted by the Producers Monthly, July 1966.

Chapter 3. The Petroleum Industry of the 1960's

Oil Shale: Its Status and Problems by J. Wade Watkins and Harry R. Johnson, Interstate Oil Compact Commission, Phoenix, Arizona, Dec. 12, 1966.

The Automobile and Air Pollution: A Program for Progress, U. S. Department of Commerce, Part 1 Summary Oct. 1967; Part 2 Subpanel Reports Dec. 1967.

Petroleum in Perspective by Harry R. Johnson, Natural Resources Journal, the University of New Mexico School of Law, January 1971.

Chapter 4. Stimulating Oil Shale Development by Leasing

Fuels Management in an Environmental Age by G. Alex Mills, Harry Perry, and Harry R. Johnson, Environmental Science and Technology, Volume 5, No. 1, January 1971.

Final Environmental Statement for the Prototype Oil Shale Leasing Program, U. S. Department of Interior, Aug. 1973 published as six volumes:

Volume I, Regional Impacts of Oil Shale Development,
Volume II, Energy Alternatives,
Volume III, Specific Impacts of Prototype Oil Shale Development,
Volume IV, Consultation and Coordination with Others,
Volume V, Letters Received During Review Process
Volume VI, Public Hearings Held During the Review Process,

Petroleum in Perspective by Harry R. Johnson, Natural Resources Journal, the University of New Mexico School of Law, January 1971.

Changing Investment Patterns of the U.S. Petroleum Industry, 1950-68, by Harry R. Johnson, Bureau of Mines Information Circular 8472, 1970.

Chapter 5. United States Responds to the Saudi Arabia Oil Embargo

Energy Reorganization Act of 1974, Public Law 93-438.

Project Independence Report by the Federal Energy Administration, Nov. 1974 published as a summary plus individual Task Force reports that included:

Coal Task Force, Thomas V. Falkie Chairman, Bureau of Mines
Oil Task Force, Vincent E. McKelvey Chairman, Geologic Survey

133

Natural Gas Task Force, Gordon Zareski, Chairman, Federal Power Commission

Facilities Task Force, King Mallory, Chairman, Department of Interior

Synthetic Fuels Task Force, S. William Gouse, Chairman, Department of Interior

Oil Shale Task Force, Harry Johnson, Chairman, Department of Interior

Geothermal Task Force, Richard J. Green, Chairman, National Science Foundation

Solar Task Force, Don Bettie, Chairman, National Science Foundation

Nuclear Task Force, Merrill J. Whitman, Chairman, Atomic Energy Commission

Synthetic Fuels Commercialization Program, by the Synfuels Interagency Task Force to the President's Energy Resource Council, Nov. 1975. The summary was published as two volumes:

Volume 1 Overview
Volume 2 Cost/benefit Analysis of Alternative Production Levels.

Chapter 6. Clash of Government Energy Plans

The Chronicles of Martin by Martin R. Adams. Published 2014 by Virtualbookworm.com Publishing Inc., P.O. Box 9949, College Station, TX 77842.

Emergency Natural Gas Act of 1977, Public Law 95-2, Feb. 2, 1977.

National Energy Program Fact Sheet on the President's Program, President Jimmy Carter, April 20, 1977.

The Politics of Mistrust by Aaron Wildavsky and Ellen Tenenbaum, SAGE Publications, Inc., 275 South Beverly Drive, Beverly Hills, CA 90212, 1981.

Market Oriented Program Planning Study, Hearing before the Subcommittee on Fossil and Nuclear Energy Research, Development, and Demonstration of the Committee on Science and Technology, U.S. House of Representatives, Ninety-fifth Congress, July 12, 1977.

Chapter 8. United States Responds to the Iran Civil War

Panic at the Pump by Meg Jacobs, Hill and Wang, a division of Farrar, Straus and Giroux, 18 West 18th Street, New Your 10011, 2016.

Synthetic Fuels Commercialization Program, by the Synfuels Interagency Task Force to the President's Energy Resource Council, Nov. 1975. The summary was published as two volumes:

> *Volume 1 Overview*
> *Volume 2 Cost/benefit Analysis of Alternative Production Levels.*

Energy Security Act, Public Law 96-294, June 30, 1980

Windfall Profit Tax Act, Public Law 96-223, April 2, 1980

Enhanced Oil Recovery by T.M. Doscher and J.A. Kostura, SPE/DOE 14881, April 1986

Outlook for Enhanced Oil Recovery by Harry R. Johnson, Bartlesville Energy Technology Center, DOE/BETC/OP-82/4, June 1982.

Bartlesville Energy Center, The Federal Government in Petroleum Research 1918-1983 by Rodney P. Carlisle, and August W. Giebelhaus, available as DE85000134 from the National Technical Information Service, U.S. Department of Commerce, Springfield, VA 22161, 1985.

Liquid Fossil Fuel Technology by Bartlesville Energy Technology Center, Quarterly Technical Progress Report, January – March 1982, DOE/BETC/QPR-82/1, 1982.

Decontrol of Crude Oil and Refined Petroleum Products, Ronald Reagan Executive Order 12287, January 28, 1981.

Incentives, Technology, and EOR at Lower Oil Prices by J.P. Brashear, A. Becker, K. Biglarbigi and R. M. Ray, J. of Petroleum Technology, Feb. 1989.

Primary and Secondary Recovery in the Sho-Vel-Tum Oilfield, Oklahoma by Harry R. Johnson, Khosrow Biglarbigi, Loren Schmidt, R. Mike Ray, and Steven C. Kyser, DOE Topical Report DOE/BC/14000-1, October 1987.

Chapter 9. Unconventional Natural Gas Production

Secondary Oil Recovery Possibilities Cow Run Sand, Burning Springs Pool, Wirt County, W. Va. By James A. Wasson, Harry R. Johnson, and Dean W. Boley, Bureau of Mines Report of Investigations 6460, 1964.

Oil and Gas in Pennsylvania by K.J. Flaherty and Thomas Flaherty III, Pennsylvania Geological Survey, Educational Series 8, 2014.

Clinton Sand Reservoir Characteristics Essential to Successful Waterflooding by Leo Schrider, Ohio Oil and Gas Association, March 1, 1968.

Surface Studies Predict Orientation of Induced Formation Fractures in Appalachian Area. Overbey, W.K., and Robert L. Rough, Presented at meeting of the Eastern District, API Division of Production, April 1968.

Natural and Induced Systems and Their Application to Petroleum Production by J. Pasini III and W.K. Overbey, Society of Petroleum Engineers paper 2565, 1969

Hydrocarbon Production and Fractures Systems by W.K. Overbey, Jr. and J. Pasini III, API Division of Production, April 1970.

Natural Gas From Eastern U.S. Shales by Leo A. Schrider, C.A. Komar, J. Pasini III and W.K. Overbey Jr., Society of Petroleum Engineers paper 6841-MS, October 1977.

DOE's Unconventional gas Research Programs, 3.1.1 Key Eastern Gas Shales Projects by U.S. Department of the Interior.

Horizontal Well to Remove Methane from Coalbeds by Joseph Pasini III and William K. Overbey, U.S. Patent 3,934,649, January 1976

Orienting Induced Fractures in Subterranean Formation by Zane Shuck, U.S. Patent 4,005,750, February 1977

Method for Controlled Directional Drilling in Subterranean Earth Formation by Zane Shuck, U.S. Patent 4,026,356, January 1977

Horizontal Drilling in Deep Austin Chalk by Petroleum Technology Transfer Council of Louisiana State University May 1997.

Horizontal Devonian Shale Wells Fractured Without Water

1. *Devonian Shale Horizontal Well: Rationale for Wellsite Selection and Well Design* by A.B. Yost II, W.K. Overbey, S.P. Salamy, C.O. Okoye, and B.S Saradji, Society of Petroleum Engineers, SPE/DOE 1640, 1987.
2. *Drilling a 2,000-ft Horizontal Well in the Devonian Shale* by A.B. Yost II, W.K. Overbey, and R.S. Carden, Society of Petroleum Engineers, SPE 16681, 1987.
3. *Analysis of Natural Fractures Observed by Borehole Video Camera in a Horizontal Well* by W.K. Overbey, L.E. Yost, and A.B. Yost II, Society of Petroleum Engineers, SPE 17760, 1987.
4. *Inducing Multiple Hydraulic Fractures from a Horizontal Wellbore* by W.K. Overbey, Jr. A.B. Yost II and D.A. Wilkins, Society of Petroleum Engineers, SPE 18249, 1988.
5. *Site Selection, Drilling, and Completion of Two Horizontal Wells in the Devonian Shales of West Virginia* by William Overbey, William Richard Carden, David Locke, Phillip Salamy, T.K. Reeves, and Harry Johnson, Final report prepared for the Department of Energy, 1992.
6. *Liquid-Free CO2/Sand Stimulations: An Overlooked Technology* by Ray Mazza, Presented at Society of Petroleum Engineers Eastern Regional Meeting, October 2001.

The Boom in Shale Gas? Credit the feds by Michael Shellenberger and Ted Nordhaus, Wall Street J., Dec. 17, 2011.

A Geologic Play Book for Trenton-Black River Appalachian Basin Exploration by Doug Patchen, West Virginia University and others from Ohio Geological Survey, Pennsylvania Geological Survey, West Virginia Geological Survey, University of Kentucky, and New York State Museum Institute. Final report prepared for the Department of Energy, 2006.

Assessment of the Potential Impacts of Hydraulic Fracturing for Oil and Gas on Drinking Water Resources (External Review Draft) by U.S. Environmental Protection Agency, Washington, DC, EPA/600/R-15/047, 2015.

Zero Discharge Water Management for Horizontal Shale Gas Well Development by Paul Ziemkiewicz, Jennifer Hause, Raymond Lovett, David Locke, Harry Johnson and Doug Patchen. Final Report prepared for Department of Energy, 2012.

More Oil, More Water by Trent Jacobs, Journal of Petroleum Technology, December 2016.

Chapter 10. Unconventional Oil Production

Strategic Significance of America's Oil Shale Resources published as two volumes: *V. 1 Assessment of Strategic Issues* and *V. 2 Oil Shale Resources Technology and Economics* by Harry R. Johnson, Peter M. Crawford, and James W. Bunger, 2004.

Potential for Oil Shale Development in the United States by Khosrow Biglarbigi, Anton Dammer, Jeremy Cusimano, and Hitesh Mohan, SPE paper 110950, 2007.

America's Oil Shale A Roadmap for Federal Decision Making by James W. Bunger, Peter M. Crawford, and Harry R. Johnson, 2004.

Energy Policy Act of 2005, Public Law 109-58

Task Force on Strategic Unconventional Fuels published as three volumes: *V. 1 Preparation Strategy, Plan, and Recommendations, V. 2 Resource Specific and Cross-cut Plans, and V. 3 Resource and Technology Profiles*, 2007.

Profiles of Companies Engaged in Domestic Oil Shale and Tar Sands Resource and Technology Development by Peter Crawford, Emily Knaus, and Harry Johnson, 2007 updated 2008.

Is Oil Shale America's Answer to Peak-Oil Challenge? by James W. Bunger, Peter M. Crawford, and Harry R. Johnson, Oil & Gas Journal, Aug 9, 2004.

In Saudi-Shale Fight, Both Claim Victory by Benoit Faucon, Alison Sider, and Georgi Kantchev, Wall Street J., Dec 16, 2016.

Chapter 11. United States Responds to the Libyan Civil War

Energy Policy and Conservation Act of 1975, Public Law 94-163.

Panic at the Pump by Meg Jacobs, Hill and Wang, a division of Farrar, Straus and Giroux, 18 West 18th Street, New Your 10011, 2016.

Energy Security Act of 1980, Public Law 96-294.

Energy Emergency Preparedness Act of 1982, Public Law 97-229.

Strategic Petroleum Reserve Crude Compatibility Study, U.S. Department of Energy, December 2005.

Strategic Petroleum Reserve Annual Report for Calendar Year 2011, U.S. Department of Energy, Nov. 2008.

Chapter 12. Completing the Journey Toward Energy Independence

Assumptions to the Annual Energy Outlook 2016 by Energy Information Agency, Table 9.3.

The Boom in Shale Gas? Credit the feds by Michael Shellenberger and Ted Nordhaus, Wall Street J., Dec. 17, 2011.